SWAROVSKI

The Magic of Crystal

SWAROVSKI
The Magic of Crystal

Text by *Vivienne Becker*

Principal Photography by John Bigelow Taylor

HARRY N. ABRAMS, INC., PUBLISHERS

Editor: RUTH A. PELTASON
Designer: ANA ROGERS

Page 2: Max Schreck. GIANT OWL. 1983
Page 7: Adi Stocker. EAGLE. 1995

Library of Congress Cataloging-in-Publication Data

Becker, Vivienne, 1953–
Swarovski: the magic of crystal / text by Vivienne Becker;
principal photography by John Bigelow Taylor.
p.   cm.
Includes index.
ISBN 0–8109–4454–5
1. Swarovski (Firm)—History. 2. Crystal glass—Austria—Wattens—History
—19th century. 3. Crystal glass—Austria—Wattens—History—20th century. I. Title.
NK5205.S9B43 1995
748.2936′42—dc20   94–31129

Published in 1995 by Harry N. Abrams, Incorporated, New York
A Times Mirror Company

Printed and bound in Italy

CONTENTS

SWAROVSKI
The Magic of Crystal

Swarovski:
The History
of a
Family
Company

Every working day of the year, in the Tyrolean mountain-rimmed town of Wattens in Austria, the family-owned company, Swarovski, produces tens of millions of crystal gemstones, each one perfect and precise in shape and color. These stones, sent all over the world to the costume jewelry, fashion, and accessories industries, are the nucleus of a remarkable output of crystal products which have made Swarovski world leaders in the field of cut crystal. Their story is one of continued dedication to perfection and innovation throughout the past one hundred years of social, political, and technological changes.

■THE BEGINNING■ It was another century, another world, when Daniel Swarovski (1862–1956) set up his own company in the tiny, sleepy village of Wattens in 1895. Taking over an abandoned factory, he set about manufacturing the crystal stones which were to dominate costume jewelry throughout the twentieth century.

Daniel Swarovski, a technical genius and a visionary, was born on October 24 in 1862 in Georgenthal, a small village in the Iser mountains of northern Bohemia, the center of the Bohemian glass industry. This region, which was part of the mighty Austro-Hungarian Empire, had been famed for its glassmaking and glasscutting, an industry which dated back to the early seventeenth century. Like most of the inhabitants of the village, Daniel Swarovski's father was a glasscutter, and also like so many other skilled craftsmen and artisans in Bohemia, he had his own workshop at home. It was there, in his father's small factory, that Daniel Swarovski served his two-year apprenticeship. He and his father produced hand-

Max Schreck. GIANT CHATON. 1990
Height 11.4 cm, diameter 18 cm (4½ x 7⅛ in.)
This "enlarged" version of Swarovski's classic jewelry stone makes a superb decorative ornament.

The home where Daniel Swarovski was born, October 24, 1862, in Georgenthal, Bohemia.

faceted stones which were fitted with copper pins and used to decorate jewelry, hair combs, and accessories. His father sold exclusively to a firm of exporters called Feix.

Bringing to this work the fresh approach and enthusiasm of youth, Daniel Swarovski came to realize that the lengthy and labor-intensive process of manual cutting was about to be outdated and superseded by an automatic process. He realized too that the handworking of glass stones was inadequate to meet the growing demands of the market. From these early days, he began to experiment with ways of mechanizing his handicraft. His first experiments were in electroplating, and through this he discovered a better method of fixing the copper pins to the stones.

In 1880, at the impressionable age of eighteen, he went to Paris, the center of the fashion world, to demonstrate his new process to an important client of Feix's. There he also had the opportunity to see the full creative potential for his stones and to view the finished product. Back in Bohemia, he continued his experiments, now centered around the construction of glass-cutting machines.

The year 1883, however, proved to be a turning point in Daniel Swarovski's life: his own ideas and experiments crystallized when he traveled to Vienna to visit the first International Electric Exhibition, widely advertised in the newspapers. There the twenty-one-year-old saw machines invented by the pioneering geniuses of technology, including Edison, Schuckert, and Siemens, and he realized on the spot that this new revolutionary form of energy was going to change industry, and the way of the world, forever. Swarovski determined to be among the first to harness its massive potential to his own industry. He saw too that opportunities for rapid expansion and innovation in Bohemia were limited and for a while he remained in Vienna to gain valuable stonecutting and setting experience.

However, it was not long before his father called him back home. The elder Swarovski had decided to move to Wiesenthal, where he set up his own business producing riveted glass jewelry, which he exported all over the world. Daniel Swarovski continued to experiment

Daniel Swarovski in his garden, c. 1940s.

ceaselessly, at one point starting a separate business with Edward Weis, his future father-in-law, making jewelry from glass sheets and, later, colored beads from faceted glass rods. At this stage too he learned that to survive he had to maneuver around the fickle tides of late nineteenth-century fashion, and so for a brief time he also worked in metal. He moved around the area from Johannesthal to Gablonz and back to Johannesthal.

Around 1891, Daniel Swarovski began to work on a crystal brilliant-cut stone, just like the diamond, which was later to develop into the classic Swarovski stone, the "chaton." Producing small crystal brilliant-cut stones was at this time a flourishing cottage industry in and around Gablonz; hundreds of workers, including children, cut, faceted, and polished the stones by hand at home in their spare time, usually after working in the fields. To increase their brilliance, the stones were coated with a thin layer of silver. In the fashion capitals of London, Vienna, Paris, and New York, diamonds were the rage, largely due to the discovery of new deposits in South Africa, as well as to the introduction of domestic electric lighting which had ushered in a widespread vogue for white, colorless jewelry. Correspondingly, the fashion for imitation diamond jewelry was growing fast, turning into an important international industry in its own right. More and more factories were starting up in Gablonz. Daniel Swarovski became concerned with improving the technique and tools for cutting and polishing these popular stones and sought desperately to adapt his budding mechanical process to their production.

At last, his device for mass-producing crystal brilliants was a success. The advantages of his invention were speed and precision; the stones, of fine quality, and at the right price, also showed a uniformity which was most desirable. In 1892, Daniel Swarovski registered his invention in Prague and applied for a patent. In his memoirs he was later to recall, "The invention of my cutting machine made a completely new manufacturing method possible, and a completely new turn was taken in the entire industry leading to an unexpected revolution. For both the stonecutters and the jewelry manufacturers, the invention opened up vast new poten-

tial for the creation of extremely beautiful, high quality and real-looking costume jewels."
Gablonz boomed and it is no exaggeration to say that this boom was closely associated with
Daniel Swarovski's invention of a highly efficient stonecutting machine.

The machine gave Swarovski the basis for his own company's specializing in mechanically cut and polished crystal stones. However he was wise enough and experienced enough
in the business to know that he must safeguard the fruits of his hard work and the advantage
he had won over his competitors in Gablonz. He made a momentous decision to move far
away from the entire area in which he had spent his life.

"There were two important reasons why we did not open our factory in Bohemia. In the

Wattens, around the turn of
the century. At this time, there
were only about 800 residents
living in the small Tyrolean
village.

first place, it was clear to us that my invention and the new production methods it made possible would have been imitated immediately in this glasscutting region of northern Bohemia. In the second place, the region lacked the hydroelectric power we needed to operate our machines."

■ A MOMENTOUS MOVE ■ Daniel Swarovski began to look around for a suitable site, both secluded and with water power, that would enable him to develop and expand his process. Eventually, he decided on an old, abandoned textile factory in Wattens, a small village with just 744 inhabitants, about fifteen kilometers east of Innsbruck in the Inn valley. Not only was the village in a serenely beautiful setting, but the mountain stream running through it provided ample water power. On October 1, 1895, just before his thirty-fourth birthday, Daniel Swarovski and his family, including the Weis family, moved from their homeland to Wattens.

The factory was operational within two weeks and business went well from the very start. In the first weeks, some two hundred packages of "chatons" (with twelve hundred stones per packet) were supplied to agents in Paris, who in turn sold the stones to England, America, and Germany. Pforzheim in Germany at this time was developing into a very important jewelry manufacturing center. By the turn of the century, as diamond jewelry was reaching a high point, Swarovski's stones, known as *pierres taillées du Tyrol,* or Tyrolean cut stones, were sought after by costume jewelry manufacturers for their superior quality, brilliance, and precision. In homage to the Alpine region that had bred this reputation, Swarovski chose as his first trademark the noble edelweiss, a beautiful and evocative symbol of the mountains that both protected and provided for him.

■ A FAMILY BUSINESS IS BORN ■ With success came expansion and improvement of the factory and working conditions. In 1900, Swarovski bought the factory he had previously been renting, spent some 24,000 guilders on acquiring extra land, and installed electric light-

The heart of the workshop,
where the various parts of the
machinery were constructed.
Photo c. 1920s

The laboratory for glassmaking experiments. Dr. Brugger,
the director of the lab, is
shown at left. Photo c. 1920s

ing to increase efficiency and productivity. He employed one hundred workers. At this point, too, Daniel Swarovski was able to begin putting into practice a plan, based on his personal deep-seated humanitarian beliefs, for the well-being of his employees. As a young man learning about his industry in various workshops and factories, he had seen the frequent abuse of power of factory owners over their employees. That experience had a powerful effect on him, instilling in him a farsighted vision of a democratic, respectful working relationship between employer and employee. Contrary to many nineteenth-century ideals, Swarovski saw that the quality of life could be improved, rather than degraded, by the machine and the success and prosperity it would engender. As he was later to explain, he saw his workers as fellow men. "We must respect them all as human beings and help them to lead fulfilled lives in dignity and contentment."

This concern with the welfare of the workers, passed down from generation to generation of Swarovskis, has been another key element in the company's steady growth and enduring success. A highly progressive and open-minded entrepreneur, Daniel Swarovski took a personal interest in his employees, encouraging them to pursue a wide range of cultural and sporting activities. He set up a social housing scheme, which continues today, so that almost one third of Swarovski's employees now live in subsidized homes and enjoy profit-linked bonus schemes.

■ TWENTIETH-CENTURY EXPANSION AND GROWTH ■ In the early years of

the twentieth century, as fine diamond jewels in exquisitely refined light and lacy platinum settings reached a peak of excellence and popularity among the elite, so imitation diamond jewelry became increasingly fashionable and sophisticated, sold alongside the real thing in fine jewelry shops. Swarovski's high-quality stones were in great demand and by 1906 the company was growing so fast that new premises and machinery were urgently needed. A larger, modern factory was built and equipped, and in the following year, 1907, a hydroelectric plant went into operation higher up the valley, providing the essential ingredient of a more plentiful and efficient supply of energy. That same year, Swarovski started to build the first housing project for employees.

Success, however, always has its price and in this case that price was the increased competition from the glassmakers and cutters of Gablonz. When his main supplier of glass decided to start cutting stones, Daniel Swarovski decided to strive for total self-sufficiency by producing his own glass. This would also give him the opportunity to improve the raw material and so perfect and control even further the quality of his product.

In 1908, Daniel Swarovski, now working with his three sons, Wilhelm, Friedrich, and Alfred, embarked on experiments in glassmaking, conducted in a specially built workshop next to their family villa in Wattens. It took three years for them to design and build their own laboratory and furnaces, and even longer to come up with the right "recipe" for crystal that was refined to a state of flawless brilliance. This was to be a vital milestone in the Swarovski story, enabling the company to reach a new, unprecedented level of mass production. By 1913, they had at their disposal all the resources needed to expand into a large-scale business. With the glassmaking now under their control, Swarovski could also produce fine colored glass, leading to an exciting range of colored stones, thereby greatly expanding their markets.

■WORLD WAR I ■ Throughout its history, Swarovski has faced and overcome various setbacks and crises triggered by far-reaching world events. The first of these crises came in 1914 when the phenomenal growth of the company was brought to an abrupt stop by the unexpected catastrophe of World War I. Employees were conscripted into the army and under threat of appropriation of machines for the armaments industry, production almost came to a standstill. Faced with the prospect of losing everything he had worked for since his youth, as well as sacrificing the security of his six hundred workers, Daniel Swarovski was forced into supplying military equipment. Only in this way was he able to maintain his work force and keep the company going until the war was at an end.

In the control room, the completed grinding wheels were examined for quality. Photo c. 1920s

The production and sale of grinding wheels steered the company through the difficult days of World War I and led to the establishment of the Tyrolit® grinding tools and abrasives. Photo c. 1920s

During the war years there was a shortage of grinding tools and abrasives, and always one to turn a crisis into an advantage, Daniel Swarovski took this opportunity to develop his own tools. In 1917, after two years of research and development, he had succeeded in making the grinding wheels used in the cutting of crystal stones. As ever, in the process Swarovski had improved their quality and before long his tools were being sought after by outside companies. In 1919, Swarovski registered the brand name Tyrolit® for their grinding tools and abrasives. Today, a global business, eighty percent of Tyrolit® production is exported to more than eighty countries over the world. Additional manufacturing centers have been set up in Spain, Italy, France, the United States, and Argentina.

■ DECADES OF DIVERSIFICATION ■ The devastation and vast social, political, and economic upheaval caused by World War I provided a valuable lesson to Swarovski, who learned that diversification was the key to surviving such world crises. Producing jewelry stones alone left the company far too much at the mercy of fashion and fluctuating economic climates. The unexpected success of the grinding wheels, for example, had pointed the way toward a much wider range of products that appealed to diverse markets and so acted as a safety net in times of trouble.

However, in business terms, times of trouble evaporated, if only temporarily, in the madness of the early 1920s. Having survived the war with his business intact, Swarovski was poised to take advantage of a great boom period in costume jewelry, during which the demand for diamanté and for beads and crystal stones of all kinds reached manic proportions.

Gabrielle Chanel had turned entrenched jewel rules upside down, making unashamedly fake jewels and pearls not only respectable but highly desirable, even essential high-fashion accessories. Her arch rival Elsa Schiaparelli, the genius of twentieth-century fashion, instigated the true *bijoux fantaisie,* injecting artistry, wit, and fantasy into jewelry design and putting the consummate seal of approval on costume jewelry for the rest of the century.

The massive changes, the turnaround in traditional values and the relayering of society brought about by the war, were ideally represented by the new respectability given to non-precious costume jewels. The fantasy and freedom of fashion jewelry, set loose from the stultifying heirloom syndrome associated with precious gems, became the perfect expression of the new liberated, working, dancing, smoking, sports-playing woman of the 1920s. For the new working woman, with money to spend on clothes and accessories, costume jewelry provided a slice of instant Hollywood-style glamour and escapism, combining fashion and fantasy. And while couture or costume jewelry had been an amusing whim of the rich during the 1920s, it became ubiquitous during the 1930s and depression years, a time when the costume jewelry industry blossomed worldwide.

The gemstones themselves were the main components of these new ornaments, and couturiers, designers, and their workshops and manufacturers turned to Swarovski for the finest quality crystal stones. Swarovski, for its part, was careful to keep in step with the latest fashions and fads decreed by Paris; during the 1920s and 1930s, the great era of couture, the company forged close working relationships with designers and couturiers, which continue today. Swarovski, then as now, developed specially shaped and colored stones for specific designers, manufacturers, and embroiderers, both influencing and responding to high fashion. At the same time, most of the popular costume jewelry in the 1920s and 1930s followed in the footsteps of fine jewelry, imitating the grand jewels of the Place Vendôme, which featured the new geometric cuts of diamonds, eloquently echoed by Swarovski's ever-evolving crystal stones. Responding to the boom, Daniel Swarovski developed newer cutting machines and more efficient processes of mass production having even greater capacity.

During this boom period in the business, Daniel Swarovski's sons became more involved in various aspects of the company, leaving the founder free to return to his beloved inventing and to pursue his plan of diversification.

Daniel Swarovski with his three sons, Wilhelm, Friedrich, and Alfred Swarovski, c. 1930s. All three continued in the family business and considerably helped expand the family enterprise.

■ GLASS REFLECTORS FOR ROAD SAFETY ■ In 1925, experiments in Wattens centered around the production of glass-reflecting elements, which brought into play Swarovski's expertise in the manufacture of precision glasscutting. It was another fifteen years before these reflecting elements appeared on the market, and not until 1950 that they were sold under the trademarked name Swareflex. Over the years, the quality and durability of Swarovski's glass reflectors, and their maximum reflecting capabilities, have proved themselves invaluable to road, rail, and sea safety worldwide. Now used in over fifty countries, they are produced with Swarovski's computer-controlled crystal production processes. The Swareflex line comprises glass reflectors for horizontal and vertical road markings for pedestrian safety and for special railway applications. The styles include pavement markers in plastic or metal, such as lane dividers and city studs, or in the use of flexible guard posts, guardrail, and wall reflectors.

■ CRYSTALS AND FASHION ■ In 1929, Daniel Swarovski focused once again on the world of fashion, which was changing fast and reshaping the markets for cut crystal stones. He had noted that a drop in sales was due to the drastic change in hair styles: the shocking new bob had replaced the long piled-up coiffures which were usually fixed with decorative, crystal-studded hair combs. Decorative shoe buckles had also become superfluous as high-cut footwear had been altered to suit the vogue for short tubular dresses. Swarovski's aim now was to move with the times and open up new markets.

In the 1920s, dance dresses, beaded and embroidered with crystals, were all the rage. Likewise, the couture giants, led by Schiaparelli, were emphasizing and developing the art of beaded embroidery, and the leading ateliers such as the Maison Lesage were turning to Swarovski for closer cooperation. In 1931, Swarovski was ready to patent yet another invention, an ornamental crystal-studded textile ribbon or band which could be applied directly onto fabrics, shoes, or indeed on to any accessories. This was to be the start of another hugely successful and creative branch of the company, known initially as the "trimmings" or "rhinestones" department, and which has continued to grow and keep pace with twentieth-century fashion.

The Swarovski patented ribbon brought fashion and crystal closer together. Before long, the new department came up with a chaton band, a ribbon of crystals set into a special plastic that also had been developed and patented by the company.

During the 1930s, the new department set about creating various trimmings for cocktail dresses, shoes, belts, accessories, bridal gowns, and cabaret costumes. "Fantasy" banding was set with stones arranged in different shapes, such as squares or flowers, on black, gold, white, or silver plastic. Colorful embroidery beads were also produced by Swarovski. From this time, the mutually creative relationship between Swarovski and fashion designers grew even stronger. Elsa Schiaparelli herself visited Swarovski's plant in Wattens to learn as much as possible about cut crystal production and to understand the possibilities for costume jewelry design. Alongside its design innovations, Swarovski continued extensive scientific research keeping one step ahead of trends so that they could respond immediately to the moods and whims of fashion.

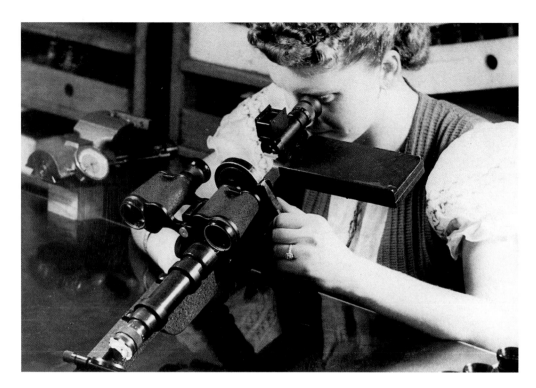

Adjusting the reticle on an early pair of Swarovski binoculars.

■OPTICAL PRODUCTS ■ In 1935, Wilhelm, the eldest son, produced his first prototype pair of binoculars, laying the foundation for yet another Swarovski product with the brand name Habicht®. The binoculars opened the door to the optical industry, which was to be a lifeline for the company during the next crisis—World War II.

■WORLD WAR II ■ In 1939, Swarovski was faced once again with the unpleasant choice between closure or accepting military orders. To safeguard the company's existence and the future of his employees, Daniel Swarovski was forced to manufacture military binoculars and optical products; in return he was able to continue a small production of crystal jewelry stones, along with that of glass reflectors and abrasives. By contrast, the costume jewelry trade was still flourishing in the United States, and stones were in great demand. The shipment of stones out of Austria to the United States was a hazardous and tricky business and usually involved an adventure following the Trans-Siberian route, by way of Vladivostok and Kyoto.

Once the war was over, however, Swarovski found itself happily at the start of a period of great prosperity and creativity in the costume jewelry industry, particularly in the United States. Swarovski became a major participant in this success, so much so that within a few years of the war the company had become Austria's largest earner of foreign currency.

Not only did Swarovski need yet more space and machinery for increased production, but in 1950, at the height of the massive jewelry boom in the United States, it found itself unable to cope with the floods of orders for stones, and a rationing system was introduced from 1951 to 1953.

■THE BOOM YEARS ■ During the war, America, cut off from European influence, had bred an entirely new jewelry style of its own—bold and effervescent, using sweeps of gold and gilt metal and large semiprecious stones. Throughout the 1940s and 1950s, this look lent itself to interpretation by the burgeoning costume jewelry industry, which had responded to the demand for high-fashion, well-made but affordable jewels and accessories for working women. The United States had none of the European preconceptions about imitation jewelry and turned costume jewelry not only into a massive industry but also into a part of American culture and an art form in its own right. The 1940s and 1950s in America, decades of gracious living, prosperity, and great style, also were the boom years of the great fashion jewelry companies such as Trifari, Napier, and Monet, and of individual designers such as Marcel Boucher and Miriam Haskell. Design had become more creatively styled, sometimes echoing precious jewelry but more often blazing a trail of its own, making the most of the fantasy and freedom offered by this expressive medium.

Swarovski was involved in the creation of many of these jewelry fashions. For example, when Trifari began to create versions of Van Cleef & Arpels's famous invisibly set jewels, they turned to Swarovski to make special stones or shapes that looked like the real thing; Swarovski in turn took ideas and designs even further than could be done with precious jewelry.

In Paris, too, costume jewelry thrived under the influence of couturiers such as Jacques Fath and Christian Dior and with the arrival of the new all-diamond jewels typical of the 1950s.

During the postwar period, Swarovski's range of colors greatly increased. Initially, the stones were produced in twenty-two colors, rising during this boom time to fifty-five and later to sixty-seven. Today, Swarovski crystal stones are produced in thousands of different shapes, colors, and finishes.

Other branches of the company were thriving too: in 1948, Swarovski Optik, specializing in the manufacture of optical instruments, was set up in the nearby village of Absam-Eichat, and two years later in 1950, Tyrolit® grinding tools and abrasives were transferred to Schwaz.

■ A STAR IS BORN ■ Meanwhile, in the prosperous 1950s, important new developments in crystal stone production were on the horizon at Swarovski. Experiments were being conducted into ways of coating the stones with thin layers of metal to further enhance their brilliance. The process involved vaporization of the metal in a vacuum. As an offshoot of these experiments, in 1955, Manfred Swarovski, the founder's grandson, working closely with Christian Dior, created the famous Aurora Borealis stone, glinting with fiery shooting rainbow colors. It was something entirely new, and the *AB,* as it came to be called, was a phenomenal success worldwide, sparking a vogue for unusual coated stones of various colors, with luminous or iridescent effects. Schiaparelli, with Swarovski, developed a particularly dramatic rocklike frosted or patinated stone in a shimmering ice blue, sea green, or deep rose red. Inspired by the success of the Aurora Borealis, Swarovski began to give their special stones evocative names such as *Jais Scarabé* or *Satinée.* Special stones were developed through the 1960s and 1970s, ranging from the *Fireball* of 1969–79, a multi-faceted crystal ball shot through with fiery

Max Schreck
LARGE PYRAMID
(VAPORIZATION EFFECT)
Two views
1976 (retired 1993)
Height 6.4 cm, depth 4.8 cm
(2½ x 1¼ in.)

colors, to the wildly successful cabochon *Moodstone* of 1976–77, which was said to change colors according to the mood of the wearer.

■THE END OF AN ERA ■ On January 23, 1956, Daniel Swarovski died at the age of ninety-four. He left behind him one of the most prestigious and productive companies in Austria, and with it a legacy of limitless possibilities for continued growth and creativity. He left too a philosophy of dedication to perfection, which has been scrupulously upheld by subsequent generations of Swarovskis.

After the death of its founder, the company continued its commitment to new inventions, new paths, and new products. In 1957, Swarovski decided to capitalize on its expertise in cutting and polishing crystal stones by applying the processes to precious gems and a special gem-cutting department was set up. Then in 1965, the company began to explore further the relationship between crystal and light by looking carefully at the lighting industry, which had in fact been using Swarovski products for many years. Swarovski started to manufacture chandelier parts and in 1977 developed the trademark Strass®. Today, the company makes hundreds of different full-cut crystal parts in all shapes and sizes for the world's leading lighting manufacturers, providing for designers an ever-changing range of new crystal elements and shapes. The unrivaled weight and brilliance of Swarovski crystal with its special radiance and light-refracting qualities is ideally suited to the making of chandeliers. The components also have a special finish that prevents dust and moisture from adhering. Swarovski crystals now hang in the most impressive chandeliers in public buildings around the world from the Metropolitan Opera House in New York to the Palace of Versailles.

The 1960s were years of rapid expansion at Swarovski as well as in the worlds of art and design. In 1967, Swarovski began to manufacture and mechanically cut synthetic jewelry stones, including the brilliant cubic zirconia, the closest imitation of a diamond, and synthetic corundum and spinel. These simulated precious stones are manufactured in a laboratory from the

Chandelier in shades of
sapphire and clear crystal,
designed by Hervé Leger
for his showroom in Paris,
with custom-made elements
specially commissioned from
Swarovski.

finest raw materials under conditions as close as possible to those of nature in an effort to

produce synthetic gems of superb quality, color, and uniformity. The company was the first

to find a way to cut cubic zirconia by mechanical methods. Marketed under the name Swaro-

gem®, the synthetic stones, as well as some natural stones such as green agate and marcasite,

are all cut and polished with immaculate precision and consistency in a wide variety of shapes

and sizes, which have proved invaluable to jewelers around the world. One of the newest devel-

opments at Swarogem® has been the production of cubic zirconia in four colors as well as

the traditional white.

It was around this time too that Swarovski tackled the technical challenge of creating

their own simulated pearls, an integral part of costume jewelry, and now an important complement to the vast range of crystal gemstones.

▪ THE TRANSFER PHENOMENON ▪

Meanwhile, in the flourishing rhinestones or trimmings department, a revolutionary new invention was on its way. After the sensational 1960s, fashion had lost its way and the 1970s were largely an age of non-ornamentation. Swarovski revived glamour and luxury by bringing crystal decoration to casual clothes such as denims and T-shirts, swimwear, hosiery, and sportswear. After years of research and technology, an exclusive transfer technique was perfected in 1972 and introduced to the industry in 1975. It was a child of Swarovski's supreme marriage of design and technology: an adhesive method for permanently fixing ready-made crystal patterns to any fabric by heat and pressure. Once fixed, the stones cannot be removed by washing, dry cleaning, or even by saltwater.

At first Swarovski simply marketed the process, but before long they were creating their own patterns and designs, and today the rhinestones department has become a creative force in the fashion and textiles trade, showing three new collections twice a year as well as working on special commissions. The range of stones used in the transfer process has grown enormously to include pearls and pearlized cabochons, metal studs, plastic foils, and much more.

▪ A NEW DIRECTION ▪

The economic crisis of the 1970s, particularly the period 1973 to 1974, adversely affected Swarovski's fortunes. While the oil crisis caused havoc with economies around the globe, social changes such as the back-to-nature and the women's movements significantly impacted fashion. Decorative costume jewelry went into decline, and sales at Swarovski suffered.

Once again, the company's management was forced to look carefully at ways in which they could outwit the crisis and in the future not fall prey to such an extent to outside factors. Following in their founder's footsteps, their answer was to move even further toward

Anton Hirzinger
MAXI SWAN. 1995
Height 16 cm, depth 16.6 cm,
width 10.6 cm (6³⁄₁₆ x 6½ x
4³⁄₁₆ in.)

complete self-sufficiency, thereby reducing their dependence on the manufacturing industries and fluctuating markets. Swarovski decided to produce their own range of finished products in addition to supplying parts to other manufacturers. It was a bold and expansive move in a new direction, involving a major reorganization of the company's structure and the setting up of an international distribution network and a substantial number of sales offices.

■ SWAROVSKI SILVER CRYSTAL ■ In early 1976, two commemorative crystal items for the Olympic Games at Innsbruck tested the market. Later in the year Swarovski's technical department came up with an unexpected breakthrough: a transparent adhesive to fix together crystal shapes. Four chandelier parts were glued together to give birth to a little mouse, the first member of Swarovski's crystal menagerie. Soon, other crystal animals followed, among them a hedgehog, tortoise, and swan, marking the debut of the successful Swarovski Silver Crystal line.

So enthusiastic was the response that the range of objects was expanded to include decorative home accessories such as candleholders and paperweights. At first all the pieces were made from existing chandelier components, but as the venture became increasingly successful full-cut crystal parts were manufactured especially for the Silver Crystal line. The collection now comprises one hundred and twenty pieces, sold all over the world.

These crystal creatures seemed to have a knack of turning delighted customers into devotees and devotees into avid collectors. Spurred on by public response and endless inquiries, letters, and requests, the company set up the Swarovski Collectors Society in 1987 to provide crystal lovers with more information about the collections, the objects, the company, and the material itself. No one could possibly have foreseen the phenomenal success of the society, nor the immense flood of applications for membership. Today the society has over two hundred thousand members in more than twenty countries around the world, with eighteen branch offices. The society offers extended customer service and advice, and special benefits

Max Schreck. MOUSE. 1976
Crystal, rhodium-plated metal. Height 2.5 cm, length 4 cm (1 x 1 9/16 in.)
*The first creature in the Silver Crystal menagerie, the mouse, simple
yet highly stylized, was assembled by hand from existing chandelier parts.*

exclusive to membership, notably the opportunity to buy annual limited editions. Over the years, special activities and events have been organized around the world to bring collectors together and there is a twice-yearly magazine, published in seven languages and edited by the SCS director Cherry Crowden.

■ A NEW AGE OF ADORNMENT ■ To mark the company's new directions and ventures, the edelweiss trademark, which had been faithfully retained for so many years, was replaced with the image of a swan: pure, white, and elegant, and a symbol of metamorphosis appropriate to Swarovski's transformation of crystal throughout the century. The depiction of the swan in the logo shows the tail studded with colored stones.

The new image marked an important turning point in the company's aims and ambitions. Swarovski embarked on a mission to turn the crystal object into an item in tune with the spirit of the age. Design and creativity became foremost considerations in the company's new philosophy, aligned, as always, with technological virtuosity. It was Swarovski's technical brilliance that brought them total design freedom and the opportunity to position crystal as one of the most exciting, versatile, and modern mediums of the late twentieth century.

Costume jewelry became the single most important fashion accessory of the 1980s. Freed by this new age of adornment from restrictions of design, manufacture, and particularly of social stigma, costume jewelry once again blossomed as an industry and as an art form, worn at all times of day, on all occasions, however formal. For Swarovski, the 1980s brought immense new scope. In 1988, the company acquired the celebrated Paris embroidery atelier Montex. Founded in 1938, Montex supplies the haute couture industry with hand embroidery and beading and the fashion industry with embroidered fabrics. Since it began, Montex has always been a major customer for Swarovski crystals, which are an intrinsic part of Montex's exquisite embroidery. Dedicated to luxury and fine workmanship, Montex works hand in hand with couturiers such as Chanel, Dior, Yves Saint Laurent, and Christian Lacroix, making catwalk and haute couture sensations as well as ornamentation for their ready-to-wear lines, and cre-

*Alessandro Mendini for
Daniel Swarovski*
CENTROTAVOLA
(CENTERPIECE). 1989
Clear and jet black crystal
Height 15.6 cm, diameter
27.3 cm (6⅛ x 10¾ in.)

*Ludwig Redl for
Swarovski Selection*
DOSE SHIVA (SHIVA JAR). 1993
Crystal, platinum-plated metal
Height 7.1 cm, diameter 10.8 cm
(2¹³⁄₁₆ x 4¼ in.)

ating decorative trimmings for accessories by houses such as Isabel Canovas, Hermès, and the new Daniel Swarovski line.

Around this time Swarovski turned its attention to the serious revitalization of the cut crystal object. In a monumental move looking ahead to the future, Swarovski commissioned a limited edition range of exclusive and avant-garde objects from leading Italian architects and designers Alessandro Mendini, Ettore Sottsass, and Stefano Ricci. The objects were part of the new Daniel Swarovski collection of couture accessories, launched in 1989 and master-minded in Paris by Rosemarie Le Gallais. According to Le Gallais, the objects, candlesticks, bowls, ashtrays, table centerpieces, and objets d'art worked well alongside the belts, jewels, and handbags. Like fashion accessories, these limited-edition pieces are today's expressions of personality and individuality. The Daniel Swarovski accessories have achieved worldwide recognition among an elite clientele due to their quality, exclusivity, and extraordinary cre-ativity. They are sold in the world's most exclusive department stores and specialty boutiques.

In January 1992, the first Daniel Swarovski boutique opened on the rue Royale in Paris. Designed by Roland Deleu, the jewelry box—like shop echoes the collection's unique blend of traditional luxury with contemporary design. The boutique is the first of a series planned for the fashion capitals of the world to provide appropriate settings for the Daniel Swarov-ski collection.

As part of this corporate move toward the manufacture and marketing of finished prod-ucts, it was vital for Swarovski to produce a range of costume jewelry bearing its name. After several years of exploration and preparation, the Swarovski Jewelers Collection emerged in department stores and high-profile boutiques around the world. The Swarovski Jewelers Col-lection successfully blends classic elegance with the latest jewelry trends, using the finest materials and superior handcrafted finishes.

In spring 1992, a new designer range of crystal objects was introduced to bring serious modern crystal design to a wider clientele. Called Swarovski Selection, the striking twelve-

Daniel Swarovski collection. EVEREST EVENING BAG
Autumn/winter 1991. Black satin, hand embroidered clear and frosted crystal
The hundreds of crystals embroidered on this bag form a dramatic icicle-like motif.

piece collection comprises vases, bowls, clocks, and other functional yet beautiful objects, designed by six carefully chosen and highly respected artists: Joël Desgrippes, Ludwig Redl, Bořek Šipek, Giampiero Maria Bodino, Adi Stocker, and Martin Szekely. Two new designs are added to the Selection line every year and new artists and designers recently drafted into the project include design guru Andrée Putman.

■ THE FUTURE ■ Daniel Swarovski saw the purity of crystal as a reminder of the constant need for purification of the environment. In keeping with the founder's concerns for the well-being of the world around him, in recent years Swarovski has been involved in a number of cultural and environmental projects. In Wattens, a hydraulic recycling plant was implemented in 1988 to clean and recycle water used in the cutting process. The amount of water purified every day is in the region of 14,000 cubic meters, equivalent to a swimming pool 2 meters deep and the size of a soccer field. In 1990, measures were also taken to reduce air pollution by connecting the company to the Tyrol natural gas supply.

From a cultural point of view, Swarovski has assumed an important, high-profile role as the sole sponsor of a series of international exhibitions closely linked to its own history. "Jewels of Fantasy," the first major exhibition of twentieth-century costume jewelry, was launched in grand style at the Museo Teatrale alla Scala in Milan in February 1991, the first venue on a worldwide tour scheduled to end in 1995, the year of Swarovski's centenary. Through 350 diverse examples of costume jewelry from Art Nouveau and Jugendstil, through Art Deco and Modernism, the Hollywood style of the 1940s, to the swinging sixties and the present day, the exhibition looked at the work of both famous designers and previously unrecognized artisans, tracing the evolution of design and costume jewelry as a cultural phenomenon.

"The Cutting Edge: 200 Years of Cut Crystal" embarked on its U.S. tour in San Francisco in 1992. Here, the exhibition looked back at the story of the cut crystal object from the eighteenth century to the present day. Running at the same time was "Imperial Austria: Treasures of Art, Arms and Armor from the State of Styria," a collection of more than 250 paintings, sculptures, engravings, swords, firearms, and suits of armor from the armory and other museums in Graz, Austria. More sponsored exhibitions are planned for the future.

■ 100 YEARS ■ The year 1995 marks the centenary of the company—the perfect time to take stock of achievements and to look ahead to the future. Swarovski, established by one extraordinary man in an abandoned factory in Wattens in 1895, is now one of Austria's largest companies, with more than eight thousand employees. The company remains privately owned and is run by the family with the same deep commitment to constant innovation, quality, and the pursuit of excellence as it was one hundred years ago. With a view to the future, Swarovski still channels a sizable portion of profits back into research and development. The parent company is located in Wattens, while administration, sales, and the Swarovski Collectors Society headquarters are based in Feldmeilen, on Lake Zurich in Switzerland.

■ A CRYSTAL WORLD ■ Every year thousands of collectors come to Wattens to visit the home of Swarovski crystal. As recently as 1988, the company realized that a center was needed to provide information, education, and entertainment for these visitors. The aim of such a center was to emphasize Swarovski's role as the world's leading manufacturer of machine-cut crystal and to explain the story of crystal as well as the company's philosophy and culture and its commitment to technological progress, innovation, and design.

At present, Swarovski is building a center in Wattens whose opening will coincide with other events during the company's centenary year. The designer of the center is André Heller, a Viennese multimedia artist. The visitors' center is actually nestled within a manmade hill so as to be integrated with the surrounding mountains and is set in a typical Alpine parkland with a small lake; the entrance facade is made to look like a large giant, who stands guard over the building. This new "crystal world," like so much of Swarovski's individual crystal objects and accessories, is part of the company's pursuit of beauty and magic in this wondrous medium.

The Story of the Crystal Object

Although glass has been highly prized as a versatile material for luxury wares since antiquity, the origins of crystal as we know it today, a particularly brilliant and lustrous form of clear glass suitable for cutting, can be traced back to eighteenth-century England. It was in the late eighteenth century that cut crystal, then in its infancy, became a treasured status symbol. In the two hundred years hence, glass has weathered vicissitudes of taste and style, to the point where, in the late twentieth century, it has undergone virtually a total transformation.

Glass was first made some 3,500 years ago, although how it actually began is still somewhat uncertain. The art of glassmaking, accompanied by that of glasscutting, was taken up, developed, and refined by the Egyptians, Persians, and Romans, and after virtually disappearing in Europe during the Dark Ages, while still thriving in Byzantium, it resurfaced in the splendor of Renaissance Venice, an age of glassmaking glory.

Around the fourteenth century, the master glassmakers of Venice came up with a colorless glass that they called *cristallo* because it looked like the mineral quartz, known as rock crystal, which was also widely used for precious and decorative objects. *Cristallo* was exported all over the known world and admired and copied for centuries. Although too thin and brittle for wheel cutting or engraving, it could be scratched with a diamond point to achieve some surface decoration. Since the time of ancient civilizations, glasscutting and engraving seem to have largely followed the path of the noble art of gem engraving, flourishing especially in the glassmaking centers of northern Europe, particularly Germany,

FOOTED HUNTING GOBLET. Bohemia, 1597
Height 27.2 cm (10¹¹⁄₁₆ in.). Collection The Corning Museum of Glass, New York
The interest in enamel ware remained active in northern Europe;
this particular goblet is typical of those made between 1590 and 1610.

CIRCUS BEAKER. Roman Empire, 3rd century AD Height 9 cm, diameter 10.4 cm (3½ x 4⅟₁₆ in.) Collection the Danish National Museum, Copenhagen *Enamel work on glass, the technique employed on this beaker, was done during the first century, in Syria, and probably in Egypt and Italy.*

Bohemia, and Silesia. It was in the early seventeenth century that Caspar Lehman, a gem engraver at the Imperial Court in Prague, first applied his wheel-engraving techniques to glass. From this time, splendid engraved glass, of magnificently intricate compositions, became the prized possessions of the rich and royal.

Like their German counterparts, Venetian glassmakers and artisans were highly regarded, ennobled, and honored, while the secrets of their glassmaking were so closely guarded that the craftsmen were often forbidden to leave the city on pain of death. It was not until the seventeenth century that some of these secret techniques were revealed to the world in a book called *L'Arte vetraria,* written by Antonio Neri, a Florentine, in 1612.

A half century later, the Englishman George Ravenscroft (1632–1683), a technician

COVERED CRISTALLO BEAKER
Venice, c. 1530
Height (with cover) 27.5 cm
(10 ¹³⁄₁₆ in.)
Collection The Corning
Museum of Glass, Corning,
New York
*The beaker actually served as a
communal drinking vessel used
at meetings of German guilds.
Diamond scratching was used
to inscribe the members of some
similar such group.*

RAVENSCROFT ROEMER
England, c. 1676–77
Height 18.8 cm (7⅜ in.)
Collection The Corning
Museum of Glass, Corning,
New York
This lead glass Roemer *bears
the seal of a raven's head, which
was Ravenscroft's own mark.*

FILIGREE-COVERED GOBLET-
STYLE VASES. Murano, late
16th century, early 17th century
Collection Staatliche Museen
zu Berlin

employed by the Glass Sellers' Company in London under the direction of the Duke of Buck-
ingham, may well have been influenced by this book, which had been translated into English
in 1662. Ravenscroft had been called into the company around the 1660s to advise and experiment, with the aim of creating a more profitable home-based industry that was not reliant
on Venetian imports. Ravenscroft experimented to change the prevailing composition of glass,
adding English flints instead of Venetian pebbles and changing the alkali from soda to potash
in an effort to come up with a truly English glass. In 1676, working secretly at a second

glasshouse in Henley-on-Thames, Ravenscroft created lead glass, a heavy, lustrous material having excellent light-reflecting qualities. Ravenscroft's addition of lead oxide to the glass resulted in an entirely new material, similar to but more durable than the Venetian *cristallo*. Among the outstanding virtues of the new lead crystal were its flexibility and its suitability for cutting and engraving. Initially, the new lead glass was often left uncut, relying on simplicity to show its brilliance, clarity, and luster to full advantage; it was not until the early to mid-eighteenth century that English glassmakers made the most of the material's refractive qualities by cutting it in facets, a German fashion probably introduced to England at the time of George I's accession to the throne in 1714.

■ AN AGE OF SPLENDOR ■ The eighteenth century was the supreme age of light, luxury, and glamour. Domestic lighting had greatly improved due to new and better wax candles that burned longer and more brightly, and this new flickering candlelight became one of the most distinctive features of the age. Candlelight shed a glow of romance and intrigue on social events, which could now take place at night. The new English prismatic crystal was the perfect partner for candlelight: faceted, jewellike drops, tumbling in waterfalls of rainbow-colored light, were fashioned into chandeliers and candelabra, far lighter and more refined than those made previously of rock crystal.

At the same time, the social scene had changed drastically in the first half of the century and by the 1850s distinctions were fading. Out of the growing industrial age came a flourishing new middle or business class, made up of self-made merchants and their families with wealth and disposable income to spend on luxuries once reserved for the aristocracy. A critic in the magazine *The World* commented in 1755 on the recent social changes saying, "We are a nation of gentry; we have no such thing as common people among us: between vanity and gin the species is utterly destroyed. . . . Every tradesman is a merchant, every merchant a gentleman, every gentleman one of the noblesse."

EPERGNE. England, c. 1760–80
Height 49.2 cm (19½ in.)
Collection The Cooper-Hewitt,
National Design Museum,
Smithsonian Institution/Art
Resource, New York. The
Bequest of Walter Phelps
Warren
*A stellar example of the impor-
tance of cut glass found in the
home of the English aristocrat.
The multiple faceting and refrac-
tive qualities suggest the unusual-
ly high level of status this object
would confer on its owner.*

Everyone who was anyone, with any pretensions to fashion and status, acquired a selection of the excitingly novel cut crystal ornaments, geared particularly toward the fashionable ritual of the dining table. As well as splendid and extravagant chandeliers, there were cruets, dishes, salvers and plates, drinking glasses, punchbowls, sweetmeat dishes laden with glistening candied fruit, towering silver-encased centerpieces, their brilliance enhanced by the soft candlelight used in the eighteenth century. This aristocratic image of the domestic interior, especially its emphasis on fine dining, has been at the heart of the cut crystal tradition ever since.

As light pervaded the entire era, the eighteenth century also became the great age of the diamond, with which cut crystal has always enjoyed a special kinship. Diamonds, discovered in Brazil about 1729, were eagerly sought after but still in relatively short supply. In Paris in the 1730s, a young and enterprising jeweler from the Quai des Orfèvres, Georges Frédéric Strass (1701–1773), sensing a gap in the market for a rival to the diamond, began making exquisitely refined jewels from English and Bohemian lead crystal. The crystal could be cut, faceted, and polished just like diamonds and precious gems, but in even more exciting shapes and sizes, and backed with tinted foil to intensify its color and sheen. Very soon these so-called strass or paste jewels became all the rage, sold in the most exclusive boutiques in Paris and worn by the highest echelons of Parisian society.

In tune with the stylistic evolution of Rococo to neoclassicism, fashions in the cutting of crystal gradually changed throughout the century, moving from flat, broad facets in the shape of diamonds, crescents, or triangles to crisp crosscut diamond patterns at the end of the century and then to typical Regency miter cutting and later to flat-flute cutting. The light and fanciful shapes of the Rococo style were replaced by the more austere linear forms of neoclassical design as the eighteenth century came to a close. In reaction to the excessive frivolity of Rococo, and encouraged by astonishing archaeological discoveries, neoclassicism looked back to classical antiquity for design motifs and forms. In England, particularly, cut crystal

PITCHER AND WASH BOWL
England, c. 1740–70
PITCHER: height 23.6 cm,
diameter 15.1 cm (9¼ x 5¹⁵⁄₁₆ in.);
BASIN: height 10.1 cm, diameter 37.1 cm (4 x 14⅝ in.)
Collection The Corning
Museum of Glass, Corning,
New York
The simplicity of these two pieces is belied by the variety of cuts, among them ovals, diamonds, scallops, and flutes.

objects became simpler, plainer, and heavier, often urn or amphora shaped, sometimes combining deep cutting and engraving in a single object.

Meanwhile, the chandelier reached a high point in style and popularity during the Regency period, about 1810–1830. The focal point in all fine interiors, the chandelier was finely proportioned and exquisitely constructed for maximum grandeur and light reflection, with as much crystal as possible covering all its parts, usually in waterfalls or tiers of crystal drops. This period marked the height of elegance, artistry, and ingenuity in English cut crystal, which was to have such a profound impact on worldwide trends. So popular was glasscutting during the early nineteenth century that the shape and form of the object itself tended to be subordinate to the decoration.

A typical trade card for a London glassworks. Courtesy of the Board of Trustees of the Victoria and Albert Museum, London

The taste for cut crystal, well established at this time, had spread as far as the United States and as nearby as Ireland. There it grew to be an important industry, notably in Waterford, which has remained a world-famous center. The Waterford glasshouse was set up by two brothers, George and William Penrose. By 1783, they were ready to advertise their wares, both useful and ornamental, in the *Dublin Evening Post*. English and Irish cut crystal, chandeliers, tableware, novelties, and luxuries were exported to the rest of Europe and particularly to the United States, where cut crystal was greatly admired. Decanters, barrel shaped or straight sided, were among the most popular items, along with bowls and vases, jugs and ewers, covered bowls and serving dishes. As demand grew and production increased, the cutting process began to be mechanized in the late eighteenth century, heralding a new age in the story of the crystal object.

Illustration of engraving, buffing, blowing, and cutting glass at the Glassworks Exhibit of Gillender & Son, Philadelphia, on the grounds of the Centennial Exposition, Philadelphia, in 1876. From Frank Leslie's *Illustrated Newspaper,* November 18, 1876. Collection The Corning Museum of Glass, Corning, New York

■ THE AGE OF EXHIBITIONS ■ By the nineteenth century, cut crystal generally became more complex and elaborate in design. Here, in the great age of consumerism, it symbolized prosperity, luxury, and technological triumph. Considering its origins in the eighteenth century, where it had been the prerogative of the nobility, the cut crystal object conferred an aura of respectability and social status, so eagerly sought after by newly wealthy middle-class Victorians.

At the start of Queen Victoria's reign, the fashion in crystal cutting and decoration had already shifted from diamond cutting and miter cutting to broad, flat-flute cutting, combining horizontal and vertical lines with diamond facets. However, by midcentury, the diamond cut had become popular once again, more glittering and crisply complicated than ever before, appealing to a broader clientele hungry for novelty and decorative home accessories. In the United States, a cottage industry in crystal cutting had developed in the 1820s, largely pioneered in glass factories around Pittsburgh and Boston. Also about this same time, the fashionable English and Irish and now American cut crystal had begun to be imitated by glass pressing, a revolutionary method which produced a distinctive lacelike effect with an opaque stippled background. The introduction of pressed glass was another major milestone in the history of the decorative glass object, making it more widely accessible since its earlier, elitist hand-crafted associations. Competition from pressed glass encouraged makers of cut crystal to be ever more inventive in terms of technical prowess.

Setting aside its social connotations, crystal was also the ideal expression of the industrial age, signifying man's supremacy over nature. The ultimate sign of crystal's importance was its prominence in the construction of the magnificent Crystal Palace, masterminded by Prince Albert to house the 1851 Great Exhibition of the Works of Industries of All Nations, a showcase for the highest artistic and technological achievements of the day, and the embodiment of Victorian taste. The building itself, designed by Joseph Paxton, drew great public attention to its spectacular use of glass. The centerpiece of the Crystal Palace was a glass fountain twenty-seven feet high, the creation of F&C Osler of Birmingham. Even the iron piping and supports were completely encased in molded and cut glass, enhancing the overall effect of the impressive structure.

Inside the exhibition as well, glass ornaments predominated, particularly British cut crystal, with displays of towering layered centerpieces and epergnes, hung with silver vases, bowls, and baskets; matching tableware sets, decanters, bowls, dressing table ornaments, candlesticks, lamps, candelabra with glass icicle attachments, chandeliers draped with crystal swags and chains. The most flamboyant and expensive creations were snapped up by Eastern and Colonial princes and potentates. British crystal creations vied for supremacy with the marvelous cut glass from Bohemia and the rest of Europe, which led to an assimilation of foreign ideas and styles. The gathering of exhibitors from ninety countries fueled the prevailing trend toward historical revivals and concocted the distinctive high Victorian style of eclecticism. In the design of cut crystal, forms were taken from ancient glass, from urns and amphorae, with decoration incorporating Egyptian, Roman, and Greek motifs, as well as Moorish and Eastern elements. The Gothic Revival style, so important in the 1840s and shown extensively at the Great Exhibition, lent itself well to cut crystal, which often incorporated diamond-cut arch-shaped panels.

The lifting of the Excise Tax on glass in 1845 had given a boost to the glass industry. Since 1745 glass had been taxed according to its weight, and now with that restriction gone it could afford, literally, to be massive and elaborate, creating the sort of ostentatious grandeur of monumental proportions that appealed to nineteenth-century tastes. Demand escalated and the industry blossomed; glasscutting workshops became bigger and busier, mechanization more sophisticated, designs more complex and innovative. However, this overenthusiastic exploration of crystal cutting aroused criticism from the aesthetic commentators of the era, who were waging a war against the excesses of Victorian opulence, which depended upon standardization of design and increased mechanization at the expense of artistry within the applied arts. Cut crystal was seen as an unnatural contortion of the original material, and complicated diamond cutting was thought to detract from the innate properties of glass, its transparency and simplicity. The cutting of crystal was, it was thought, intended to make glass look like something it could never be, like rock crystal, a pretension that was abhorrent to the art critics. Furthermore, its distinctly diamondlike sparkle, so beloved of the international nineteenth-century bourgeoisie from Prague to Pittsburgh, was seen to be meretricious. John Ruskin, the theorist, art critic, and leading exponent of the new aesthetic movement, famously condemned all cut crystal as "barbarous, for the cutting conceals its ductility and confuses it with crystal."

■ ART NOUVEAU AND ART GLASS ■ The new design direction that emerged in the last quarter of the nineteenth century advocated a return to fluid, sinuous natural forms, seen as the inspiration and basis for all good decorative design. In this sense, cut crystal could never emulate nature, nor take on fluid, organic forms. Art Nouveau, the free-flowing decorative style of the turn of the century, was truly a revolution—short lived, intense, and aimed at revitalizing and unifying the degraded applied or so-called minor arts, which included glass. Now artists and designers from all fields and backgrounds, notably Emile Gallé in France and Louis Comfort Tiffany in the United States, turned their attention to glass, raising it to

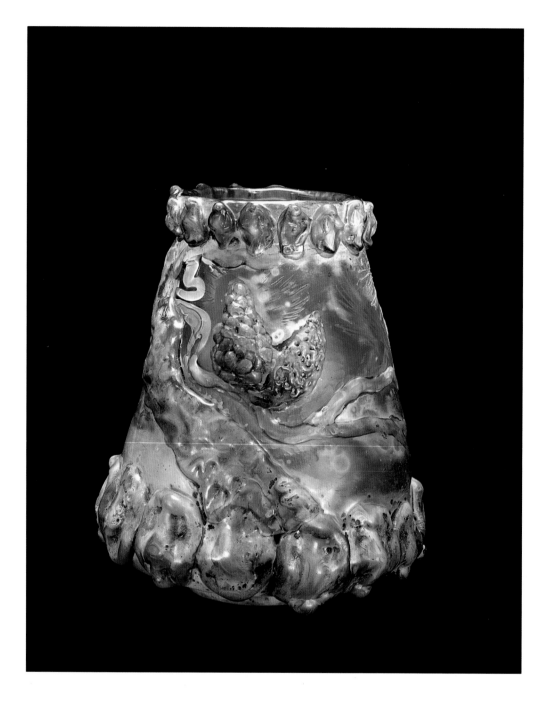

Emile Gallé
PINES. 1903
Height 17.8 cm (7 in.)
Collection The Corning
Museum of Glass, Corning,
New York. Purchased with
funds from the Clara Peck
Endowment, the Houghton
Endowment, and a special
grant
*Among various practitioners
of the Art Nouveau style, the
Frenchman Emile Gallé
displayed a virtuoso talent
and imagination.*

Opposite:
Louis Comfort Tiffany
FLOWERFORM VASE
Tiffany Studios, c. 1900–5
Height 36.2 cm, base of diame-
ter 16.5 cm (14¼ x 6½ in.)
Private collection
*One of Tiffany's more delicate
workings in favrile glass.*

WHEEL-CUT PLATE. United States, c. 1906
Diameter 34 cm (13⅜ in.). Collection The Corning Museum of Glass, Corning, New York. Gift of T. G. Hawkes & Co.
By the middle of the 19th century, incisions in cut glass had reached almost manic proportions, both in England
and in the United States. This dish is remarkable for its display of a "Russian pattern," patented by
T. G. Hawkes & Co. in 1882, and later adapted for a cut-glass service ordered by the White House in 1885.

EWER. United States,
c. 1852–63
Height 30.8 cm (12⅛ in.)
Collection The Corning
Museum of Glass, Corning,
New York
*Made at the Long Island Flint
Glass Works of Christian
Dorflinger in Brooklyn, this is
a particularly refined example
of blown and cut glass.*

the level of an art form in its own right. Colored, layered with patinas, cameo carved, shaped into spectacularly amorphous, vegetal and organic forms, art glass, as it was known, came alive with all the artistic energy of the fin de siècle. In America, the importance of art glass began to challenge the supremacy of cut crystal. Cut crystal represented traditionalism and social and material success; art glass was modern, and stood for free thinking and artistic appreciation.

In the same quarters, however, Art Nouveau had its detractors. In more formal environments, cut crystal was still in vogue, especially in England, where it reached a new height of popularity in the 1880s and 1890s, when cutting was as elaborate as technology allowed, with the emphasis firmly on mechanical marvels. Difficult rounded shapes, such as a basket, could be cleverly covered with mathematically precise patterns. Other favorite items included the square-shaped whisky decanter with heavy ball stopper and assorted toilet bottles with silver tops.

In America, cut crystal persisted as a mark of social prestige. The American "brilliant" period of cut crystal straddled the centuries and lasted until the social upheaval caused by World War I. Rich, deep-cut decoration in a wide variety of motifs, including fan shapes, miters, and stars tended to cover the entire object, achieving a diamondlike brilliance which was much admired. In many ways, the sparkling complexity of glass at this period was a last ditch attempt to recapture the safety and social elitism of the eighteenth century, and to cling to the leisured living and social order of the Belle Epoque.

■ MODERNISM ■

Art glass had brought the first winds of change in the perception of the glass object as an art form, but crystal, inextricably associated with prismatic cutting, was omitted from the process of revitalization in favor of softer, more malleable glass.

René Lalique, master jeweler and the acknowledged master glassmaker of the early twentieth century, created his own *demi-cristal,* a cross between the two, furthering the metamorphosis of the decorative crystal object.

René Lalique
MASK. France, 1928
Satin crystal. Height 31 cm
(12⅜ in.)
Courtesy Lalique, Paris
Among the most unusual and splendid works by Lalique at this time.

Art Glass, like Art Nouveau, was part of the quest to create worthy, honest objects of artistic integrity. As the new art movement swept its way across Europe, it took on different characteristics, absorbing the ideas of artists, designers, and architects in different countries. The Bohemian glass industry, which during the nineteenth century had been influenced by English and Irish cut crystal, began to have a more assertive style of its own, influenced by ideas emanating mainly from Vienna and also from Czechoslovakia itself. In Vienna, modernism manifested itself in a streamlined, architectural, linear style, stripped of unnecessary ornament, which arose from a rejection of historicism and florid surface decoration. This geometric, streamlined look was well suited to cut crystal.

The Wiener Werkstätte, founded in 1903 and based in Vienna, was a commune of workshops founded and run by a group of artists and designers who had broken away from the traditional art establishment and were committed to the unity of the arts, to a return to the traditions and quality of handicrafts. Backed by the young banker Fritz Waerndorfer and led by artists Josef Hoffmann and Koloman Moser, the Wiener Werkstätte excelled in metalwork, leatherwares, wood, ceramic, and glass. Since they could not perform actual glass manufacturing, their designs were executed by outside Bohemian glass firms. Otto Prutscher, for example, a talented all-round designer and one of Hoffmann's students, designed a range of glassware, vases, and decanters made by Bakalowits and Sons of Vienna in 1907. Prutscher's creations have become celebrated examples of the Wiener Werkstätte style and helped bring cut crystal into the twentieth century. The long stems of Prutscher's goblets incorporate the now-famous checkerboard motif, in a column of blue and clear glass squares, and a bowl, with broad fluting, is also edged with cubelike motifs. A little later, around 1911 to 1914, Hoffmann designed a range of glass for the Viennese glassmakers J. & L. Lobmeyr, who also employed geometric motifs and lines, within the traditional framework and forms of vases, toiletry sets, and drinking glasses. Hoffmann also designed for the German firm of Moser, which came to prominence in the 1920s and 1930s.

While this was happening in Vienna, Czech artists were working with the Bohemian glass industry, notably the architect Jan Kotera who came up with strong modern glass designs, simple and architectural, making a clear break with the past. In 1902, he designed a punch bowl and glasses, which were exhibited the following year at the 1904 World's Fair in St. Louis. It was simple and streamlined, yet fluid and classic in form, with broad flat cutting. Kotera was the leader of a group of Czech Cubists who used their native glass industry to express their ideas of increasingly streamlined abstraction. Pavel Hlava, a Czech glass designer, is perhaps best known for his cut and engraved glass. It began a tradition that has continued throughout the twentieth century.

Important progress was also being made in other countries with a tradition in cut crystal. In Scandinavia, the modern movement, which reached a high point in the 1950s, was characterized by clean, sweeping sculptural lines derived from natural, organic forms. Around 1915, the Swedish firm Orrefors hired artists Simon Gate and Edvin Hald, who were heralded for their work in modern glass design. In England, New Zealand–born architect Keith Murray, a functionalist and modernist, was employed by the firm of Stevens and Williams in 1932 for three months of the year to improve and update the standards of English glass design, which was still extremely conservative, particularly in the area of cut crystal. Murray was interested in old glass and impressed by the Swedish, Viennese, and Czech designs he had seen at the Exposition des Arts Décoratifs et Industriels Modernes in Paris in 1925. His aim was to revive the simplicity of old English glass, with flat, rather than deep, cutting. In the 1930s, Clyde Farquharson designed new glass for John Walsh and also for Stevens and Williams.

In the United States, Steuben Glass, established in 1903 by the Stourbridge glassworker Frederick Carder, had developed its own special lead crystal in the 1930s. Highly refractive, with great purity and brilliance, it was originally intended as an optical material for lenses but was soon being used for decorative glass. In 1935 Steuben, which had been bought by Corning Glass Works from Carder in 1918, took the step of commissioning artists and design-

Sidney Waugh for Steuben Glass. GAZELLE BOWL. United States, 1935
Copper-wheel engraved, blown clear crystal, resting on a solid crystal base cut into four flanges.
Diameter 16.5 cm (6½ in.). Collection Steuben Glass, New York
With its frieze of 12 leaping gazelles, this bowl is one of
Sidney Waugh's—and Steuben's—most classic and stunning works.

Emile Wirtz for Daum. VASE. France, 1925
Blown glass with acid technique. Height 25.7 cm (10⅛ in.) Collection Daum
The typical Art Deco design is seen in this French vase.

Edvin Ohrstrom for Orrefors
ARIEL VASE. Sweden,
c. 1925–40
Heavy clear glass with blown
pattern. Height 16 cm (6⁹⁄₁₆ in.)
The Toledo Museum of Art,
Ohio. Gift of Mrs. Hugh J.
Smith. Jr., 1948

ers to produce the widest possible range of crystal ornaments, a tradition which has contin-

ued to the present day.

During the 1920s and 1930s, particularly in France and Belgium, another important impe-

tus for cut crystal came from the world of the decorative arts—the world of luxury goods—

rather than from the more architecturally inspired and conceptual modernism. The new Art

Deco style, which had been brewing for many years and which reached a climax in 1925, was

applied to glass and crystal production as it was to all areas of the decorative arts. Art Deco

had much more to do with surface decoration, with ornamentation and style, and in a sense

formed a bridge between academic design experiments and popular taste. In most cases, the traditional wares of the cut crystal trade—toilet bottles, decanters, vases—took on the iconography of Art Deco with its dramatic geometric motifs of frozen movement and stylized figural themes. Purely ornamental, nonfunctional cut crystal and glass sculpture also began to be appreciated during this period. Baccarat in France and Val Saint Lambert in Belgium specialized in Art Deco cut crystal, which was shown to the world in Paris in 1925.

In the late 1960s and 1970s, during the general crafts revival, glass went through yet another incarnation as artists and designers working in the modernist ethic sought to re-create the ideal of art glass. As was the case in previous decades during the century, cut crystal was largely neglected, although struggles to adapt it as a more modern medium still continued.

Meanwhile, in Austria, Swarovski had been perfecting their product, honing the technology of crystal cutting and thereby creating the climate for an entirely new approach to cut crystal objects. Their technology was to provide the freedom of design that had been lacking during the earlier part of the century. Until the 1970s, Swarovski had been involved only with supplying elements—cut crystal jewelry stones, chandelier parts, fabric trimmings—to various industries. In the 1970s and even more so in the 1980s, they began to look seriously at the overall image of cut crystal. Their line of crystal animal sculptures had achieved a phenomenal success, clearly answering a worldwide desire for beautiful, ornamental, and highly personal objects made of fine faceted crystal. Now it was time to reposition the cut crystal object in the world of art and design.

With modernism well and truly interred in the archives of the 1970s, the postmodern style and sensibilities provided the perfect aesthetic package, with its deep interest in and awareness of the past, and its intention of bringing cultural references, particularly from the eighteenth century, into today's modern world.

Claudia Schneiderbauer. BUTTERFLY ON LEAF. 1994
Clear and matte crystal. Height 5.8 cm, width 4.2 cm (2⁵⁄₁₆ x 1¹¹⁄₁₆ in.)
A new interpretation of one of the most popular of all Silver Crystal emblems,
this complex design shows the growing trend toward greater naturalism in the line.

Max Schreck. BIRDBATH. 1980
Clear and matte crystal. Height 6.1 cm, diameter 10.8 cm (2⁷⁄₁₆ x 4¼ in.)

MODERN
CRYSTAL
CLASSICS

In 1976, Max Schreck, one of Swarovski's most experienced technicians, was playing with chandelier parts and a new transparent glue when he unwittingly conjured up a magical creation that was to add a new chapter to the story of cut crystal and bring pleasure to millions of people around the world. That magical creation—a mouse—and subsequent other Swarovski animals, among them the hedgehog, tortoise, and swan, showed how cut crystal could be used to depict the natural world. Without realizing it, Max Schreck bonded together two highly emotive elements: the animal kingdom and the realm of crystal. The Swarovski Silver Crystal range (so-called because of its silvery brilliance) has developed and matured enormously since those first experimental days. Today, the objects are sought-after collectors' items, highly sophisticated both in terms of design

and technology. They are inspired by carefully researched themes having a global appeal and contemporary message. Swarovski crystal animals and objects have become modern classics.

■ GLASS ANIMALS IN HISTORY ■ Animals have been portrayed in glass for some 3,500 years, dating to the time when glass was first invented. Whether man's closest friends or deadliest foes, loved or feared, venerated or vilified, animals have always been imbued with certain magical or supernatural powers and have been depicted in everyday objects and works of art of all ages. At different times in history these portrayals have shown various stylistic tendencies, naturalistic, highly stylized, or anthropomorphic; the best known example of representation in our time is the ubiquitous teddy bear. The Silver Crystal

Max Schreck. LARGE SWAN. 1977
Height 6 cm, length 6.3 cm (2⅜ x 2½ in.)
One of the earliest creatures, the swan's simple composition is still effective today.

range encompasses all three of these trends.

By about 300 BC, the ancient Egyptians were incorporating such animal motifs as the ram, frog, falcon, and jackal into their glassware. Precious objects such as glassware were buried with the dead, and portrayals of animals, some of them Egyptian deities, were intended to soothe and protect the owner's path into the next world. Roman glass was often ornamented with animal motifs, but these were mostly inspired directly by nature. Horses, lions, pelicans, and marauding mounted warriors swarmed in rhythmic, stylized patterns over Islamic glass, while Venetian Renaissance masterpieces were often painted with the wonders of the natural world, including an array of exotic birds such as peacocks. The trend continued in the glass centers of sixteenth- and seventeenth-century Germany and Bohemia, where favorite motifs included hunting scenes, stags, rabbits, and eagles. Occasionally the whole of the glass object, usually a bottle or spirits flask, took the form of the animal itself. By the industrial eighteenth century, these natural themes were largely replaced by purely abstract decorative glass designs. During this time, the newly formulated medium of cut crystal developed its own decorative language, which was not especially well suited to representational forms. Although the natural world continued to inspire nineteenth- and twentieth-century glass artists—particularly the art glass exponents—the trend did not spill over into the cut glass industry, which followed the path of geometric, prismatic, and occasionally modernist or sculptural designs.

Certainly the animal as decorative object, be it bronze, ceramics, glass, or gemstone, came into its own at the turn of the century. In Russia, Fabergé's repertoire of exquisite jeweled toys included a captivating range of carved hardstone miniature animals, brilliantly observed and interpreted to capture the character and essence of each creature, some entirely naturalistic, others highly stylized. They ranged from the humblest mouse or pig to the most curious species, such as the monkey or hippopotamus. An ambitious series of Fabergé figures, commissioned by King Edward VII, was based on animals from the royal farm at Sandringham; today these figures remain in the royal collections.

Around the same time, the Renaissance tradition of bronze sculptures enjoyed a huge revival in Europe in the late nineteenth century in the form of the brilliantly detailed *animalier* sculptures, a specialty of both French and Viennese sculptors. After World War I, purely ornamental animal objects were created in glass and crystal, but not strictly speaking in full cut crystal, which was still regarded as a medium largely incompatible with renditions of nature. Major glass artists and manufacturers of the twentieth century turned their talents to animal themes. In France, Lalique's sophisticated vases, bottles, lamps and lighting, and car mascots were all fashioned from their creator's beloved animal world. Other firms such as Daum in Nancy, founded around 1870 in the wake of the art glass movement, and Baccarat, the aristocratic glassworks set up in the eighteenth century, launched a series of sculptured crystal animals around the late 1950s and more significantly in the 1960s; Baccarat's line was designed by glass artist Georges Chevalier.

In Italy, vivacious colored glass animals thrived in the 1950s, notably at the famous Venetian firm of Venini, while in the United States, the tradition had been taken up by Frederick Carder in the late 1920s and, under his direction, later fully explored by the Steuben Glass company (which had taken over Carder's firm). In the 1970s, crystal animal sculptures, with some cutting, had become an integral part of the company's output.

■THE SILVER CRYSTAL SENSATION ■ Meanwhile, at Swarovski in autumn 1976, the first mischievous mouse in full-cut crystal was being produced and sent out into the world, blissfully unaware of its long lineage. Initially, production methods were more in the nature of experimentation: the chandelier parts—in this case a round ball, a square faceted base, and two pear-shaped drops glued together—were finished by hand, the rhodium-plated brass whiskers were individually trimmed, and the little tails were each cut out from aluminum foil. The mouse was highly stylized but also endearing and appealing in its innocent simplicity, with its air of startled alertness. A week before Christmas 1976, a limited edition of a thousand was offered for sale to employees of Swarovski in Wattens. They sold out almost immediately.

In spring 1977, the mouse was joined by a spiky crystal hedgehog, closely followed by a lovable tortoise and elegant swan, whose streamlined composition worked especially well in crystal. Tiny technological works of art, these animals, gemlike and tactile, managed to capture the personalities of the real creatures, with a particularly brilliant sparkle achieved by the juxtaposition of so many differently shaped elements and forms of facets, packed closely together, and reflecting light from so many planes, inward and outward. They were greeted so enthusiastically by the public that within a short period of time the crystal menagerie grew substantially. At first, the animals were still made from existing chandelier parts, but before long, as demand grew and the huge potential of the series became evident, full-cut components were manufactured exclusively for the animal range. At first, too, they were composed and constructed by Swarovski technicians, but soon special designers were drafted to invent and interpret new ideas. Alongside the animals home accessories were created, such as the water lily candleholders and geometric paperweights. Silver Crystal was in full swing.

Max Schreck, creator of many original Silver Crystal classics, was a devoted and highly skilled Swarovski designer-craftsman. He was born and brought up in Wattens, where his father had worked alongside Daniel Swarovski as a mechanical engineer since 1909. As a child Max showed an aptitude for art, but following his father's advice he too trained as a technical engineer and later joined the Swarovski company. Max Schreck has now retired and today there are seven resident designers: Adi Stocker, Gabriele Stamey, Michael Stamey, Martin Zendron, Anton Hirzinger, Claudia Schneiderbauer, and Edith Maier.

■THE STORY OF CREATION ■ Each of the team of designers has been specially trained at the College of Glassmaking and Design at Kramsach in the Tyrolean lowlands. The college, set up in 1947, is unique in providing training courses that are not available anywhere else. The courses deal with all aspects of the glassmaking industry, manufacturing techniques, design, finishing, and decorating. The original glassworks set up in Kramsach in 1627 had

Max Schreck
MEDIUM HEDGEHOG. 1985
Clear and jet black crystal
Height 5 cm (2 in.)
A more mature relation of the
original hedgehog and part of
the Summer Meadow group.

Max Schreck
LARGE TORTOISE. 1977
Height 4.5 cm, length 5 cm
(1¾ x 2 in.)
The group of Endangered
Species shows how the Silver
Crystal themes are in tune with
prevailing international concerns.

remained operational until 1936. After the Second World War, lack of expert workers prevented the reestablishment of the works, but as there was a huge demand for training by young people a school was set up instead. During their four-year course there, Swarovski designers have learned every practical aspect of their trade, concentrating on maximizing the potential for light and beauty in crystal.

Their challenge now at Swarovski is to create an animal figure that is as lifelike, or full of life, as possible, bringing out the most instantly recognizable and lovable characteristics of each creature. The emphasis is very much on naturalism, as well as on the legends, associations, and stories, age-old or modern, behind each animal. The designers are given a subject briefing usually with extensive documentation in the form of books, articles, photographs, monographs, and videos. Occasionally rough sketches are done, but usually the designer will start to model directly in Plasticine, or even in crystal itself, hand cutting and polishing to produce the desired effect. Very often this preparatory work is done on a much larger scale than that of the final object in order to show in greater detail the facets and modeling.

Once an initial model has been made, it is handed to the technicians whose job it is to turn the "blueprint" into the actual object, developing and perfecting a prototype for production. The technicians, usually highly skilled with a formal education in engineering, have to craft the tools which are used to make the basic shapes of the design, known as blanks. This part of the process is painstaking, complex, and time-consuming, and the designers collaborate closely with the engineers to come up with the correct tooling for each part of an animal. Although on average there are some ten elements to each object, there can be as many as fifty or sixty in a particularly complex design.

During the production process, the molten crystal is poured into metal molds to create the blanks, or pieces of solid uncut crystal. The crystal blanks have to be annealed, a process of heating and cooling in a special oven known as a *lehr* that strengthens the crystal and elim-

Martin Zendron
WHITE STALLION. 1993
Height 10.7 cm, length 8.2 cm
(4¼ x 3¼ in.)
This ambitious study in movement and action was inspired by the Lippizaner horses of the Spanish Riding School in Vienna.

Adi Stocker
MOTHER GOOSE. 1993
Clear and matte crystal
Height 6.5 cm, width 4.15 cm
(2⁹⁄₁₆ x 1⅝ in.)
*From the Barnyard Friends
group, this Mother Goose leads
her amusing little family of baby
goslings, Tom, Dick, and Harry
(not shown).*

Gabriele Stamey
MINIATURE CHICKS. 1988
Height 2 cm, width 1.5 cm
(⅝ x ¹³⁄₁₆ in.)
*Conjuring up a joyful image of
new life, these tiny chicks from
the Barnyard Friends group,
busily pecking at grain, look and
feel like tiny amuletic treasures.*

inates internal stress that might cause it to crack. The largest blanks can take up to one week to cool. At each stage of the production process the components, like the finished products, are subject to the most stringent quality controls. The blanks are graded for clarity, just like diamonds, and once declared internally flawless they are ready for cutting with optical precision grinding wheels, followed by the final processes of smoothing and polishing.

Given the continual emphasis on quality, the long path of research and development is strewn with hurdles, both technical and aesthetic. One of the major ongoing challenges lies in using crystal to suggest the fluidity of nature. Movement and energy also have to be conveyed in this blend of creativity and technology. For each object, from first briefing to finished product, the development time is usually two years.

■ THEMES AND INSPIRATIONS ■ The first animals produced in the late 1970s were selected by a somewhat random process, but today's Silver Crystal creatures are based on carefully researched themes. Ideas may take years to nurture. Sometimes the rendition of a particular animal or idea is meant to be an authentic representation, other times it may be amusing, childlike, or dreamlike. Each object, however, must produce that all-important instant emotional response that lies at the heart of the success story of Silver Crystal. Over the years themes have touched on any number of common interests and animals; today, the Silver Crystal line includes Barnyard Friends, with their miniature chicks and family of geese, Pets' Corner, such as dogs and cats, the architectural Silver Crystal City, a Game of Kings Chess Set, When We Were Young crystal nursery "toys," Crystal Melodies musical instruments, and the luscious harvest of Sparkling Fruit.

Since its inception in 1987, the Swarovski Collectors Society has also issued an annual limited edition series of three objects, organized around a specific theme or message. Each edition, limited by a cut-off date after which the tools are destroyed, is available only to members of the society, one per member. The first of these, inspired by the motivation for the society,

Max Schreck
THE LOVEBIRDS. 1987
Clear and matte crystal
Height 10 cm, width 6.5 cm
(3¹⁵⁄₁₆ x 2⁹⁄₁₆ in.)
*Caring and Sharing provided
the theme for the first Swarovski
Collectors Society annual edition,
launched in the society's inaugur-
al year, to an overwhelmingly
enthusiastic response.*

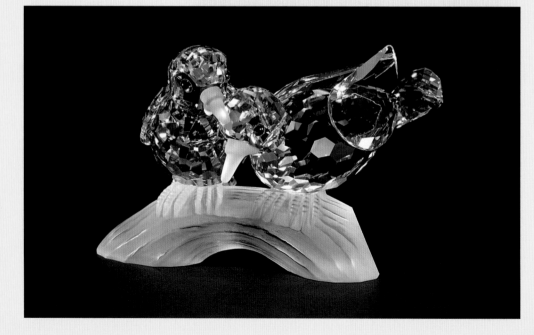

Adi Stocker
THE TURTLEDOVES. 1989
Clear and matte crystal
Height 8.5 cm, length 6.7 cm,
depth 10 cm (2⅜ x 3⅜ x ⅜ in.)
*An amorous finale to the initial
Caring and Sharing series of
SCS annual editions, now closed.*

Adi Stocker
THE WOODPECKERS. 1988
Clear and matte crystal
Height 10.9 cm, width 6.65 cm
(4 5/16 x 2 5/8 in.)
*A particularly popular pair in
the Caring and Sharing series of
SCS annual editions, now closed.*

Michael Stamey
THE WHALES. 1992
Height 10 cm, length 9.3 cm,
depth 7 cm (4⅝ x 3¹¹⁄₁₆ x 2¾ in.)
*These massive legendary creatures
glide over a crystal wave; the last
in the Mother and Child series of
SCS annual editions, now closed.*

Michael Stamey
THE DOLPHINS. 1990
Height 7.8 cm, length 12.6 cm,
depth 7 cm (3⅟₁₆ x 5 x 2¾ in.)
*With the dolphins came an
important breakthrough in design
and manufacture, since maximum
movement and naturalism were
incorporated in this model.
Inspired by the theme Mother and
Child, this piece is the first in the
second series of Swarovski Collectors Society annual editions, now
closed.*

Michael Stamey
THE SEALS. 1991
Height 5.3 cm, length 9.4 cm,
depth 7 cm (2⅛ x 3¹¹⁄₁₆ x 2¾ in.)
*Crystal perfectly conveys the icy
seascape that provides the natural
habitat of this Mother and Child,
the second in this series of SCS
annual editions, now closed.*

Gabriele Stamey. SANTA MARIA. 1991
Clear and matte crystal. Height 9.3 cm, length 11.4 cm (3⅝ x 4½ in.)
One of the most dramatic pieces in the series When We Were Young, *the stately ship in full sail is redolent of the happy, carefree days of childhood.*

Gabriele Stamey
OLDTIMER. 1989
Height 3 cm, length 8 cm
(1³⁄₁₆ x 3³⁄₁₆ in.)
A classic car, reminiscent of the early 20th century, from the enchanted nursery world of When We Were Young.

was the lovebirds, the first of the Caring and Sharing series. The lovebirds, emblems of togetherness, called *Les Inséparables* in French, signified a bond for life, a loving commitment between two creatures. Exquisitely modeled and cut in the finest detail, the lovebirds proved an enormous success and were followed by the woodpeckers, a sculpture showing a parent bird feeding its young, and then by the amorous turtledoves. Each of these devoted pairs of birds was perched on a satin-finished matte crystal branch, twig, or tree. The next series, called Mother and Child, was inspired by the interest in wildlife conservation and the international heightened sensibility about endangered species. The dolphins, designed by Michael Stamey for 1990, captured the imagination of crystal lovers everywhere, while at the same time marking a vital breakthrough in Swarovski's development of these crystal objects. The design of this piece was full of movement and life, leading the way toward ever more naturalistic creations. The dolphins were followed by the seals and then the whales, all much-loved marine mammals

in danger of extinction. From 1993 to 1995, the elephant, kudu, and lion—all limited editions—
have revolved around the theme Inspiration Africa, again tuning into the new caring attitude
of the 1990s and the universal concern for the preservation of life.

■THE SYMBOLISM OF FLORA AND FAUNA ■ Many of Swarovski's modern classics
incorporate and reflect the complex system of symbols and imagery long associated with animals. The much-beloved frog, for example, has always been considered a magical creature.
Generally recognized as a symbol of transformation, for the Egyptians the frog represented
fertility and creativity, whereas for the Romans it was an emblem of wedded bliss and as such
was revived as a favorite sentimental figure in the nineteenth century. For Swarovski, the frog,

Michael Stamey. THE ROSE. 1993.
Length 8.1 cm (3⁹⁄₁₆ in.)
A perennial message of love, the rose is a symbol of true love and the perfection of nature.

part of their Beauties of the Lake group, has been a tremendous success, clearly indicating its appeal among collectors of crystal. Likewise, the butterfly, the symbol of the soul, with its ethereal beauty and allusion to metamorphosis, has proved one of the most popular creatures in the Silver Crystal range. The swan, also a classic symbol of transformation and of purity and elegance, is much in demand, while the families of little creatures, like the miniature chicks, who suggest new life, are irresistible, and can be handled and played with like rich, glinting jewels.

That universal emblem of true love, the rose, has likewise received a generous response from collectors. In this case, the portrayal of the flower was that of a perfect specimen: an expression of ideal beauty in its prime, with delicate sparkling dewdrops on the leaves and tightly folded petals depicted by a clever mixture of modeling for the innermost blossom and cutting for the outer petals. Like so many other examples in the collection, the rose makes a perfect gift, an eloquent messenger of love and affection.

■THE ROMANCE OF CRYSTAL ■ Silver Crystal animals, flowers, insects, and objects, romantic and radiant, have clearly answered a growing need for both security and escapism in today's fast-moving life. There is a strong desire too for more magic in the world around us, and for small, amuletic or ritualistic objects, rich in meaning and symbolism. At the same time, there is a marked return to traditional values and, in our highly technological world, a longing for familiar or traditional objects that suggest comfort and continuity. Throughout literature and history, crystal, with its curious mix of transparency and durability, its union of spirit and matter, its light and luminosity, its mysterious inner life and fire, has come to symbolize clarity, and more grandly still, the truth. Its light-refractive qualities stand for optimism and happiness. In combination with more historical or archetypal associations of the animal world, the Swarovski crystal pieces have become for many cherished objects, deeply personal yet utterly universal, having immediate appeal.

Claudia Schneiderbauer. HUMMINGBIRD. 1992
Clear and matte crystal. Height 7 cm, width 6.5 cm (2¾ x 2⁹⁄₁₆ in.)

Claudia Schneiderbauer. BUMBLEBEE. 1992
Clear and matte crystal. Height 5.5 cm, length 4.6 cm (2⁵⁄₁₆ x 1¹³⁄₁₆ in.)

Team design. LARGE BUTTERFLY. 1982
Crystal, gold-plated metal. Height 5.5 cm, width 5 cm (2⅛ x 2 in.)
Symbol of the soul, the butterfly from the In a Summer Meadow group has proved to be a bestseller.

Faceted crystal exerts a special fascination; everyone understands the joy and excitement of seeing a rainbow of dancing light through a crystal prism and this childlike joy is embodied in these purely decorative, contemplative objects. The play of light has been a vital factor in the popularity of the Silver Crystal creatures: not only do they shower a sunlit room with colored reflections, but the sparkle from endless facets turns the objects into precious dreamlike treasures. Through these miniature technological marvels of the modern world, the myths of centuries are reflected and perpetuated. Silver Crystal objects are also affordable luxuries, the perfect gifts and small indulgences or personal rewards, with no purpose other than to give pleasure, which is the first rule of luxury. For serious devotees, a collection of crystal animals quite simply makes one's life happier.

■COLLECTING ■ The Swarovski Collectors Society provides comprehensive customer service and a meeting point for collectors. News is spread through the twice-yearly Collector magazine, edited by SCS director Cherry Crowden, and regular trips, seminars, and meetings are planned, partly as education and partly to bring collectors together. Further advantages include access to the annual limited editions, special repair shop facilities, and advice and tips on caring for the collections. One of the main achievements of the society has been to add a vital human element to crystal so that collectors understand the personalities behind the crystal creations—the work, care, and devotion to detail that goes into each piece. The designers themselves have become well known through the society, almost achieving cult status among the members. They often make personal appearances at society events to sign their creations and have been known to work their way through gatherings taking several hours.

The collectors' items, particularly the limited edition models and the retired Silver Crystal pieces, are passionately sought after, giving rise to a flourishing secondary market in which past pieces can be bought and sold. Collectors in various countries have formed their own net-

Max Schreck
EGG. 1979 (retired 1992)
Length 4.6 cm, diameter 6.3
cm (1¹³⁄₁₆ x 2½ in.)

Michael Stamey
SOUTH SEA SHELL. 1991
(retired 1994)
Height 4.7 cm, length 7.2 cm
(1⅞ x 2⅞ in.)

Anton Hirzinger. Centenary Swan. 1995
Height 5.2 cm, depth 5.2 cm (2⅟₁₆ x 2⅟₁₆ in.)
Stylized, stately, and sophisticated, Swarovski's swan commemorating their 100th
anniversary has wings scattered with crystal jewelry stones suggesting drops of water.

Adi Stocker. THE LION. 1995
Height 7.4 cm, length 12.9 cm, depth 7 cm (2⅞ x 5⅟₁₆ x 2¾ in.)
The last in the series of three Inspiration Africa animals, the handsome lion
is composed of a mixture of modeling for the head and full cutting for the powerful body.

works for finding buyers and objects, providing information on prices and availability, and coordinating buying and selling.

■ CENTENARY CELEBRATIONS ■ The year 1995 marks the one hundredth anniversary of the founding of the Swarovski company by Daniel Swarovski in Wattens in 1895. Such a historic milestone in the story of the company and of cut crystal itself is great cause for celebration and Swarovski has devised several special objects to mark the occasion.

The *Centenary Swan,* a glittering incarnation of the company logo, is a special edition open to the general public but limited to the year of issue. The design of the swan marks a new departure in Silver Crystal designs: it is far more stylized than current crystal swans in production, being more sophisticated and streamlined, and its wings are scattered with faceted jewelry stones to give the impression of sunlight glistening on drops of water clinging to the pristine feathers. The annual edition for members for 1995 is appropriately the lion, king of the jungle, last of the Inspiration Africa series, naturalistically portrayed with a mixture of modeling for the head and facet cutting for the lithe contours of the body. Another special offering in the centenary year is the eagle: a new breed of supremely naturalistic large crystal sculpture, which will be the first numbered edition, limited to ten thousand pieces. Designed with the utmost realism and drama by Adi Stocker, the beak and claws are made of sterling silver, and the Swarovski swan logo and the model number are laser-engraved inside the sculpture, deeply embedded in the crystal.

On the eve of Swarovski's anniversary, the company has taken an amusing look back at the auspicious start of the Silver Crystal collectors' items. Three of the original creations, the mouse, cat, and hedgehog, have been reproduced as charming miniature replicas, now called Three Old Friends. The first chapter of the Swarovski collectors' story comes full circle and a new chapter opens. The little mouse was mighty indeed.

Adi Stocker. EAGLE. 1995
Crystal, rhodium-plated sterling silver. Height 22 cm, width 15 cm (8⅝ x 5¹⁵⁄₁₆ in.)
The first of its kind, this impressive bird of prey is a new,
naturalistic maxi-sculpture, made in a limited numbered edition of 10,000.

Adi Stocker
BEAGLE PLAYING. 1993
Height 3 cm, width 3.4 cm
(1³⁄₁₆ x 1⅜ in.)

MINIATURE DACHSHUND. 1987
Height 2.7 cm, length 5.3 cm
(1¹⁄₁₆ x 2⅛ in.)

Michael Stamey
CAT SITTING. 1991
Height 4.5 cm, length 4 cm
(1¾ x 1⁹⁄₁₆ in.)

Opposite:
Max Schreck
GIANT OWL. 1983
Height 16.5 cm, length 10.2 cm
(6½ x 4 in.)

Adi Stocker. AEROPLANE. 1990
Height 4 cm, length 7 cm (1⁹⁄₁₆ x 2¾ in.)

Gabriele Stamey
CATHEDRAL. 1990
(retired 1994)
Height 5.7 cm, length 3.4 cm
(2¼ x 1⅜ in.)

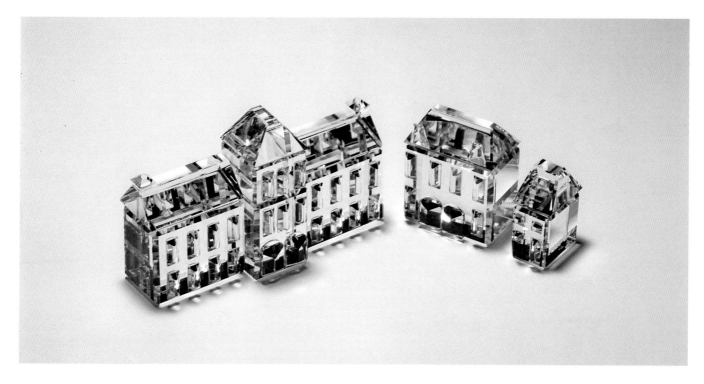

Gabriele Stamey. TOWN HALL. 1993 (retired 1994). Height 3.8 cm, length 6 cm (1½ x 2⅜ in.)

SET OF HOUSES (number 2). 1990 (retired 1994). Height 3.4 cm, length 2.5 cm (1⅜ x 1 in.)
The Silver Crystal City group provided a toy-town fantasy. Collectors could plan and build their dream town with no borders to their imaginations.

Gabriele Stamey
TRAIN SET. 1988–93
LOCOMOTIVE. 1988. Height
3.5 cm, length 6.5 cm, width
2.9 cm (1⅜ x 2⁹⁄₁₆ x 1⅛ in.);
TENDER. 1988. Height 2.6 cm,
length 2.6 cm (1¹⁄₁₆ x 1¹⁄₁₆ in.);
TIPPING WAGON. 1993.
Height 2.9 cm, length 3.9 cm
(1⅛ x 1½ in.); PETROL WAGON.
1990. Height 2.9 cm, length 3.9
cm (1⅛ x 1½ in.)
An escape back to the delights of
childhood, the parts of this train—
a locomotive, tender, wagon, tip-
ping wagon, and petrol wagon
—can be collected separately.

Max Schreck. LARGE PYRAMID. 1976 (retired 1993)
Height: 6.4 cm, depth 4.8 cm (2½ x 1⅞ in.)

Opposite:
Team design. LARGE STAR CANDLEHOLDER. 1987
Height 11.25 cm, length 14 cm (4⁷⁄₁₆ x 5½ in.)

Michael Stamey. GIANT DUCK. 1989
Height 11 cm, length 24.3 cm (4⅜ x 9⁹⁄₁₆ in.)
An impressively large creature from the Beauties of the Lake group.

Michael Stamey
SNAIL. 1986
Height 3.6 cm, length 4.3 cm
(1⁷⁄₁₆ x 1¹¹⁄₁₆ in.)

Max Schreck
LARGE PIG. 1984
Crystal, rhodium-plated metal
Height 5 cm (2 in.)

Martin Zendron
LUTE. 1992
Height 7.9 cm, width 3.2 cm
(3⅛ x 1¼ in.)

Opposite:
Martin Zendron
HARP. 1992
Height 9.98 cm, width 4.96 cm
(3⅞ x 1¹⁵⁄₁₆ in.)
*The most romantic and shapely
of traditional musical instru-
ments was chosen for the Crystal
Melodies theme.*

Michael Stamey
SEA HORSE. 1993
Clear and matte crystal
Height 8 cm, length 3.85 cm
(3³⁄₁₆ x 1⅛ in.)

Michael Stamey
THREE SOUTH SEA FISH. *1993*
Clear and matte crystal
Height 5 cm, length 8 cm
(2 x 3⁵⁄₁₆ in.)

Michael Stamey
SHELL WITH PEARL. *1988*
Crystal, simulated pearl
Height 4.8 cm, length 5.88 cm
(1⅞ x 2⁵⁄₁₆ in.)
*A perennial best-seller from the
South Sea group, the oyster shell
with its hidden treasure hints at
the mysterious enchantment of the
sea and fabled legends surrounding the pearl.*

Adi Stocker
MOTHER AND BABY PANDA
1994
Clear and jet black crystal
MOTHER: height 4 cm, length
4.5 cm, width 4.5 cm (1⁹⁄₁₆ x 1¾
x 1¾ in.); BABY: height 1.8 cm,
length 2 cm, width 1.4 cm
(¹¹⁄₁₆ x ¹³⁄₁₆ x ⁹⁄₁₆ in.).

Adi Stocker
POLAR BEAR. 1986
Height 4.5 cm, length 8.8 cm
(1¾ x 3½ in.)

Max Schreck
LARGE PENGUIN. 1984
Height 8.5 cm (3⅜ in.)
*A lovable creature from Swarov-
ski's Kingdom of Ice and Snow.*

Michael Stamey. THE KUDU. 1994
Clear and matte crystal. Height 10 cm, length 10 cm, depth 7.5 cm (3¹⁵⁄₁₆ x 3¹⁵⁄₁₆ x 2¹⁵⁄₁₆ in.)
A SCS annual edition, now closed, the magnificent kudu from the antelope family
with its towering spiral horns was the second in the Inspiration Africa series.

Opposite:
Martin Zendron. THE ELEPHANT. 1993
Clear and matte crystal. Height 8.5 cm, length 11.8 cm, width 8.8 cm (3⅜ x 4⅝ x 3½ in.)
First in the third series of SCS annual editions, Inspiration Africa (now a closed edition), this lumbering
elephant marked a technological breakthrough in the design and construction of the realistic, slow-moving legs.

THE
METAMORPHOSIS
OF CRYSTAL

The metamorphosis of crystal has begun. A fact that will surprise nobody because, after all, it is perfectly natural that a material of this caliber should have its place at the highest level of creation. What more and more people will be asking themselves is: "How is it possible that crystal has been neglected in the creative arts for so long?"
– ALESSANDRO MENDINI

One of the most significant forces in the development of twentieth century decorative design has been the cult of the individual designer. From the time of the late nineteenth century art movements, the hand of the artist-craftsman became widely recognized and valued. So important was this trend in the early twentieth century that manufacturers in a variety of industries sought out artists to rejuvenate and modernize their wares. Glassmakers called on talented designers to inject a vital contemporary note into the form and feel of their objects, aiming to make them part of the total design for living. Although glass in general has gone through many changes since Art Nouveau and the art glass movement of the 1890s, cut crystal had somehow been left behind in the wake of the nineteenth century. What was fashionable then came to be regarded over the years as bourgeois and old fashioned. Saddled with this stigma, cut crystal never properly emerged from this image with its pedigree and elegant qualities intact until quite recently.

In the late 1980s, Swarovski, at the forefront of crystal cutting technology and by this time heavily involved with marketing finished products, set about redressing the balance. Single-handed, the company has brought about a renewed interest in crystal and in doing so has changed the universal perception of cut crystal. Swarovski's idea was to commission a group of the world's most inventive and creative designers to produce an exclusive collection of limited edition, museum-quality objects. Their aim was to open the eyes of the art world to the vast possibilities of crystal and reposition the crystal object firmly in the forefront of modern design. Happily, Swarovski's high-

Daniel Swarovski collection. MAYERLING GAUNTLET GLOVES. Autumn/winter 1989
Black suede kidskin, silk lining, hand embroidered
The crystal embroidery on these gloves from the first Daniel Swarovski collection was inspired by the Viennese
Secession in homage to the company's Austrian origins and to its progressive attitude toward modern design.

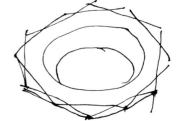

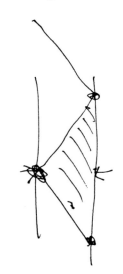

ly ambitious project, conceived in 1986, coincided with the rise of postmodernism: designers in general had rejected dogmatic functional modernism in favor of a more eclectic all-embracing approach. This new approach was particularly strong in Italy. Ever since the country emerged from World War II with a mission to reconstruct its industries and its artistic image, it has bred and nurtured some of the world's most innovative and bold designers. Collectively, they have changed the face of industrial design—furniture, everyday objects including household appliances, and materials such as ceramics and plastics have been part of this metamorphosis. Although Italian architects and artists had pushed forward the frontiers of cutting-edge design, they had overlooked cut crystal which, it must be said, was not truly part of their culture. So it was decided in Wattens, Austria, that if such designers were not coming to look at crystal then Swarovski was going to take crystal to them.

Alessandro Mendini, Ettore Sottsass, and Stefano Ricci created among them thirteen objects, candlesticks, bowls, ashtrays, centerpieces, and objets d'art. Made in limited editions of between one hundred and one thousand pieces depending on size, complexity, and cost, each was marked with the designer's signature and identifying serial number. All three design-

Ettore Sottsass for Daniel Swarovski. Centrotavola (centerpiece). 1989
Clear and matte crystal. Height 22.1 cm, diameter 25.5 cm (8¹¹⁄₁₆ x 10¹⁄₁₆ in.)

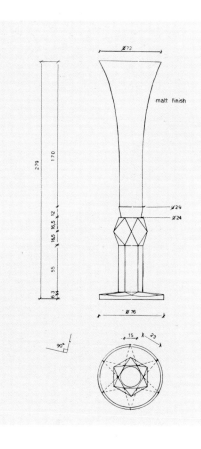

*Ettore Sottsass for
Daniel Swarovski*
Technical drawing for
Vaso Piccolo. 1988

*Ettore Sottsass for
Daniel Swarovski*
Vaso Piccolo (SMALL VASE)
1989
Clear and matte crystal
Height 31 cm, diameter 7.6 cm
(12³⁄₁₆ x 3 in.)

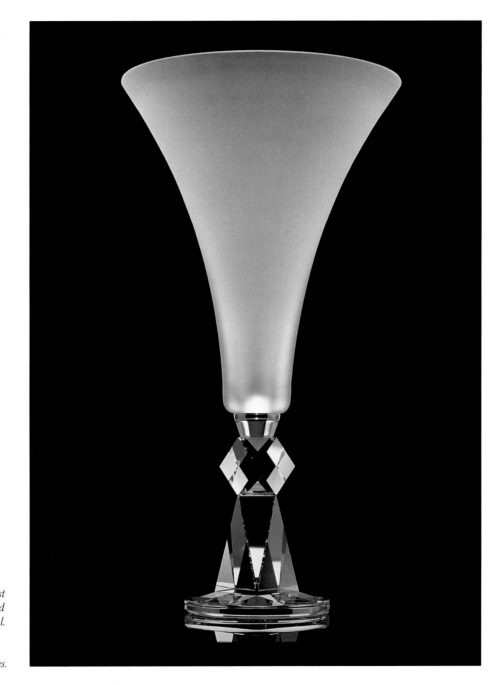

*Ettore Sottsass for
Daniel Swarovski*
VASO GRANDE
(LARGE VASE). 1989
Clear and matte crystal
Height 32.6 cm, diameter 18
cm (12⅞ x 7⅛ in.)
*Sottsass made use of the contrast
between clear faceted crystal and
smooth opaque, or matte, crystal.
Also recalling past glories of
crystal, this vase is inspired by
Venetian Renaissance glasswares.*

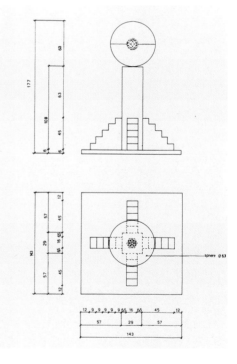

*Ettore Sottsass for
Daniel Swarovski*
OGGETTO D'ARTE
(OBJET D'ART)
Clear and red crystal. Height
27.8, width 18.5 cm (10¹⁵⁄₁₆ x
7⁵⁄₁₆ in.)
*A templelike object of contempla-
tion with an Aztec theme, the
altar is surmounted with a sphere
in which is embedded a red "eye."*

*Ettore Sottsass for
Daniel Swarovski*
Technical drawing for
OGGETTO D'ARTE. 1988

*Ettore Sottsass for
Daniel Swarovski
CANDELIERE
(CANDLEHOLDER). 1989
Height 22 cm, width 8 cm
(8⅝ x 3³⁄₁₆ in.)
The candlestick with its complex
arrangement of facets proved the
most technically demanding for
Swarovski's engineers.*

ers, each with a protean talent and a deep commitment to the future, had trained originally as architects; the pure architectural lines of their dignified objects, with their superbly balanced proportions producing crisp shimmering reflections, succeeded in releasing the full potential of the material and in showing to perfection the technical precision of the manufacturing process. This meeting of two very different cultures—Austria and Italy—and disciplines—glassmaking and architecture—resulted in a unique approach to the genre.

Embarking on the project in the late 1980s, all three designers were more than ready to turn away from 1970s modernism in favor of the newer—and more liberating—arena of postmodernism. Each designer, according to his personal interests, undertook painstaking research into the history of crystal objects and their meaning to society, and into the complex processes of manufacture, although Swarovski closely guards the more technical details of production. The architectural background of the designers suited crystal perfectly since by its very nature a crystal object is constructed from separate elements much like a building is. As with a building, too, the use of light is crucial; the individual parts of each crystal construction are joined using a special glue, pioneered by Swarovski, which does not impede the transparency of the medium.

Not surprising, either, were the highly individual approaches among the designers. Ettore Sottsass (b. 1917) has been the single most important figure in the design world since the 1970s. Sottsass was born in Innsbruck, the son of an architect from Trentino, and he too graduated as an architect, in the modernist tradition, from the Polytechnic of Turin in 1939.

Following a stint in the army, he set up his own design studio, also working on furniture, ceramics, and glass objects. In 1958, Sottsass became design consultant to the new electronics division of Olivetti and went on to design for several other industrial companies. In 1961, he visited India, which profoundly influenced his work, causing the introduction of a softer, more sensuous and spiritual tone into modern design. During the 1970s, Sottsass developed his own distinctive colorful style, inspired by 1950s design and influenced by American Pop

Art. The year 1981 saw the emergence of the controversial anti-design group, Memphis, of which Sottsass was a leader. Four years later, in 1985, he resigned from Memphis and returned to architecture.

Sottsass brought to the Swarovski project his particular interest in the ritualistic and mystical aspects of crystal. His dramatic objet d'art, *Oggetto d'Arte,* with its stepped Aztec base, also reminiscent of cathedral steps, is in fact a shrine or a temple. The "altar," or central column, is surmounted with a sphere, a symbol of eternity and contemplation, in the heart of which is encased a red hypnotic "eye." Sottsass's six designs, notably the centerpiece and large and small vases, emphasize the versatility and flexibility of the material and its surprising warmth by exploring the dramatic visual contrasts between clear and opaque matte crystal, and between clean faceted geometric forms and soft fluid contours. The vase shapes subtly echo Renaissance Venetian glass forms. Sottsass's ashtray and candlesticks provided Swarovski with technical challenges, as they maximized the effect of facets and planes that strike deep into the heart of crystal.

Alessandro Mendini's main point of reference for this project was the style and taste of the eighteenth century. The architect, who was born in Milan in 1931, concentrated on architectural journalism after 1970. He is celebrated worldwide for his radical industrial design for such companies as Alessi, Renault, and Zanussi, as well as for his furniture, most particularly the startling "Proust" armchair, completely covered with a Seurat-inspired pattern. Mendini took a more art-historical, intellectual approach to his objects, adding grandeur and richness to sleek modern lines. He delved into the eighteenth century, the Age of Reason, and spent a great deal of time in Vienna looking at objects from that period, better understanding Swarovski's cultural heritage. Baroque and neoclassical features were then simplified and distilled down to their very essence so that his resolutely modern objects retain a powerfully suggestive hint of their aristocratic antecedents. His Trophy centerpiece takes the form of a neoclassical urn, its superb proportions set off to perfection by facets recalling the

Alessandro Mendini for
Daniel Swarovski
Sketch for goblet
c. 1988

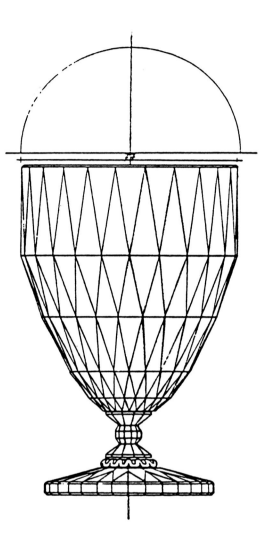

Alessandro Mendini for
Daniel Swarovski
Technical drawing for
CENTROTAVOLA. 1988

Alessandro Mendini for
Daniel Swarovski
Technical drawing for
OGGETTO D'ARTE. 1988

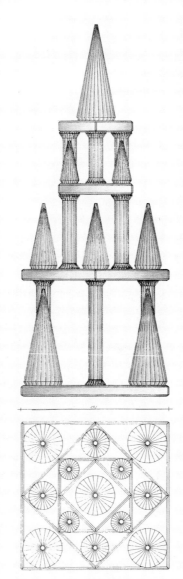

TROFEO
PER SWAROVSKI
PROGETTO NEOCLASSICO "R. ADAM"
ALESSANDRO MENDINI CON M. CHRISTINA HAMEL
25.X.88
SCALA 1:1

Opposite:
Alessandro Mendini for
Daniel Swarovski
OGGETTO D'ARTE
(OBJET D'ART). 1989
Clear and jet black crystal
Height 46.1 cm, diameter 18
cm (18⅛ x 7⅛ in.)
A sensational late 20th century
homage to the chandelier, the
quintessential aristocratic cut
crystal object. Mendini has
turned the classic cascades of
crystal drops upside down to
form a soaring ornament.

*Alessandro Mendini for
Daniel Swarovski*
POSACENERE (ASHTRAY). 1989
Clear and jet black crystal
Height 8.2 cm, diameter 13.2
cm (3¼ x 5⁵⁄₁₆ in.)

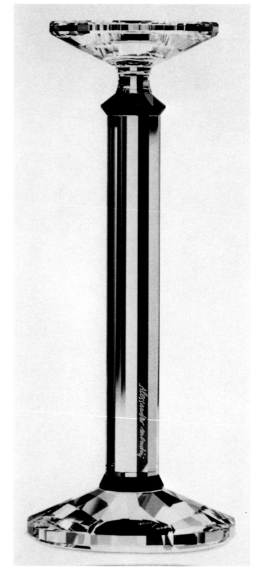

*Alessandro Mendini for
Daniel Swarovski*
CANDELIERE
(CANDLEHOLDER). 1989
Clear and jet black crystal
Height 19.9 cm, width 8 cm
(7⅞ x 3³⁄₁₆ in.)

Alessandro Mendini for
Daniel Swarovski
Centrotavola Trofeo
(trophy centerpiece). 1989
Clear and jet black crystal
Height 20 cm, diameter 13 cm
(7⅞ x 5⅛ in.)
All of Mendini's designs for
Daniel Swarovski, including
this urn-shaped centerpiece, were
inspired by late 18th century
neoclassicism, a glorious age of
cut crystal.

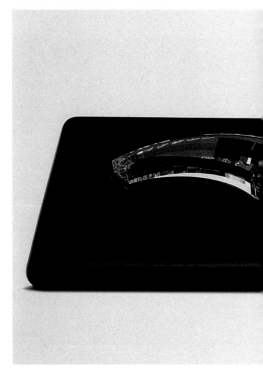

*Stefano Ricci for
Daniel Swarovski*
CANDELIERE
(CANDLEHOLDER). 1989
Height 17.1 cm, width 12 cm
(6¾ x 4¾ in.)

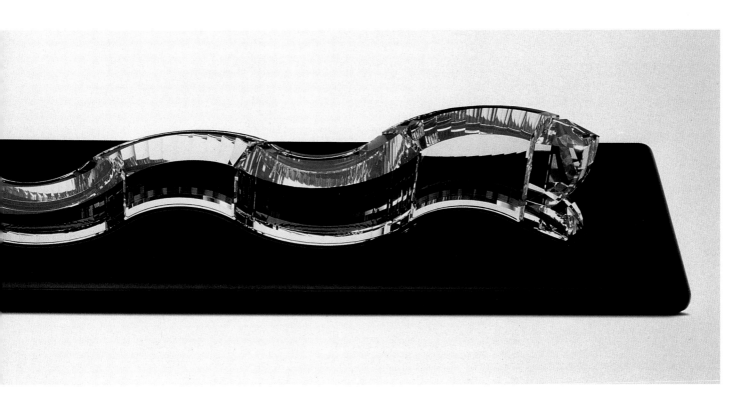

splendor of eighteenth century cut crystal. His candlesticks are neoclassically inspired columns, while his mesmerizing objet d'art wittily recalls the chandelier, the quintessential cut crystal object. Turning the cascading chandelier upside down, Mendini has created a soaring architectural tower, a powerful link between past and present.

Stefano Ricci (b. 1950), a native of Rome and an architecture graduate, is best known as a designer of jewelry for Bulgari from the late 1970s until the early 1980s. After working for a time in the area of display, packaging, and interior design, he returned to jewelry design. Ricci brought a voluptuous and sensuously decorative quality to his two objects, playing freely with the warm reflections of the crystal and their joyful rainbow effects. His candlestick cleverly blends a confident, soft roundness with long fluted facets, Regency style, while his objet

Stefano Ricci for
Daniel Swarovski
Oggetto d'Arte
(objet d'art). 1989
Height 25 cm, length 135 cm
(9⅞ x 53¼ in.)
Ricci, a respected jewelry design-
er, introduced a voluptuous,
especially tactile note into this
serpentine object.

d'art is a modern crystal sculpture, an organic serpentine stream of near-liquid light.

For all of the designers, the development of these extraordinary objects took some three years, a period of education and learning for both themselves and for Swarovski, whose technical expertise was also pushed to new limits by the artistic demands of the architects. One of the major contributions of the project to the story of cut crystal is the fact that virtually for the first time the role of technology was equal to that of the artistry. The designers were completely in tune with technology, reveling in the mechanization that is such an integral part of Swarovski's history and the history of cut crystal.

A group of these objects formed part of a Swarovski-sponsored traveling exhibition in 1992 in the United States called "The Cutting Edge," a survey of the cut crystal object during the past two hundred years. These thirteen complex modern works of art have brought the cut crystal object very much to life again in the late twentieth century. By celebrating the past and present glories of cut crystal, the three Italian architects not only restored and renewed cut crystal's position as a noble luxury but they also created modern-day objects that capture the essence of our time.

These limited-edition objects were in fact part of an even broader design venture, a range of couture accessories called the Daniel Swarovski collection, which was launched in its entirety during the haute couture collections in Paris in summer 1989. Within the framework of Swarovski's new philosophy for a metamorphosis of crystal, the objects were seen as accessories for the home, just as the fashion accessories were viewed as beautiful objects and works of art.

The Daniel Swarovski fashion accessories were as progressive and ambitious as the artists' objects. The original aim of the collection was not only to work toward a total transformation of the image of crystal but also to produce unique accessories for today's international woman. Bags, belts, gloves, jewelry, umbrellas—all exploring the glorious versatility of crystal—were designed to convert a modern wardrobe from day to night, from functional to fantastic. They are accessories of great individuality and creativity, exquisitely made to the highest standards, bringing together traditional craftsmanship and high technology. At the same time, the collection was intended to bring crystal into the forefront of design and high fashion.

The Daniel Swarovski collection represents a personal approach to high fashion, a deliberate move away from the homogenized, designer-initialed accessories of the past decade. The Daniel Swarovski line gives a designer complete artistic control over his or her ideas from start to finish, without compromising on creativity or quality. Rejecting dictatorial fashion and status symbols, these dramatic accessories herald a return to the tradition of exclusive, custom-made luxury.

Each season some of the crystal stones for each collection are specially created by Swarovski in Austria to fit a particular theme or design. The Daniel Swarovski designers also have at their disposal an outstanding archival treasure house of existing or old crystal stones from past years, which can be reinstated and used in an entirely fresh way. No expense is spared in the research, technology, and making of new tools, nor in the constant search for unusual materials, traditional handicrafts, ateliers, artists, and manufacturers.

The Daniel Swarovski collection is masterminded in Paris by Rosemarie Le Gallais. Deeply immersed in the worlds of fashion and design, she worked with Karl Lagerfeld for fifteen years. Rosemarie Le Gallais has always taken a special interest in accessories and their role in high fashion, with careful attention to their growth in the marketplace. Over the years her work had brought her into contact with Swarovski, whose crystal stones she used in many different ways. She came to realize the huge potential of this age-old yet modern material and admired the dedicated family business that has been the world's leading producer of crystal for one hundred years.

When she started the Daniel Swarovski line, Le Gallais's challenge was to express in crystal the fashion mood of the moment, along with any important cultural and social trends, such as the interest in ecology. Above all, she aimed to restore the identity and importance of accessories, which over the last two decades had become lost in a market saturated with status symbols. The nineties, she feels, are related to identity and individuality and the grow-

Opposite:
Daniel Swarovski collection
BAVIÈRE BELT
Autumn/winter 1989
Black suede kidskin, sterling
silver buckle, hand-embroi-
dered crystal mosaic
*When worn on a simple black
dress, the design of this double
belt produces a trompe l'oeil
effect.*

Daniel Swarovski collection
"IN TIME" EVENING BAG IN
THE FORM OF A POCKET
WATCH. Autumn/winter 1993
Black satin, hand embroidered
*Brilliant crystal stones and gold
threads form an intricate embroi-
dery on this "pocket watch"
evening bag.*

Daniel Swarovski collection
IMPERIAL EVENING BAG,
JARDINIÈRE LARGE AND
SMALL BROOCHES.
Autumn/winter 1993
BAG: black velvet, hand
embroidered with gold thread,
multicolored crystal stones;
BROOCHES: gold-plated ham-
mered metal, pâte de verre
leaves, multicolored crystal
stones
*The embroidered bag and
matching pair of brooches are
ornamented with crystal blooms
to form a golden jardinière.*

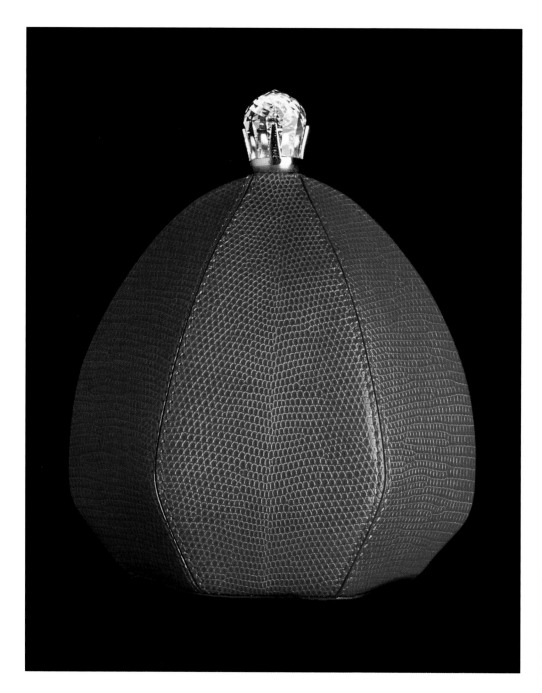

Daniel Swarovski collection
DIAMANT EVENING BAG
Spring/summer 1991
Red lizard skin, crystal clasp
*This sculpted bag crowned with
a large single crystal has become
a classic in the Daniel Swarovski
repertoire.*

Daniel Swarovski collection
"Handle" Evening Hand-
bag. Spring/summer 1993
Yellow satin, crystal mosaic
*This technique of building up
crystals into a mosaic handle was
created for the Daniel Swarovski
collection as an entirely new
way of using crystal stones as
accessories.*

ing desire to know that there is an individual behind a particular design or skill. "The nineties are years of the truth," she says. "A label alone is no longer enough, and accessories today are not simply a question of fashion, they are part of our lifestyle." Her concept was based on the realization that while women's lives were becoming faster they were also becoming more complicated so that clothes, and dressing up in general, had to become much simpler. It is mainly through accessories that women can express their personalities. Le Gallais also sensed a strong need today for a return to traditional craftsmanship and values.

According to Rosemarie Le Gallais, this is the right moment for an awakening of crystal. More and more, crystal is making a statement in its own right and has gained new value in the fashion world. Crystal is no longer considered an imitative material but is being accepted and admired for its own particular qualities. The possibilities of crystal are endless, says Ms. Le Gallais. Luminous, full of life, strong yet ethereal, brilliant yet soft, and full of mystery, crystal has become a symbol of our age.

Having at first collaborated with the designer Hervé Leger, Rosemarie Le Gallais now works closely with a team of young designers and artisans who are adept at interpreting her ideas. Each season Le Gallais comes up with a series of ideas and themes, usually starting with handbags to set the scene. Colors and materials are worked out with her team and new ways of using crystals are devised and discussed with Wattens.

The first Daniel Swarovski collection of 1989 made a strong impact in graphic contrasts of clear crystal and black, with appropriate overtones of Austrian turn-of-the-century Secessionist design, in order to link the collection with Swarovski's origins. Since that time, themes have been highly diverse, playful yet elegant, in tune with fashion yet entirely original; they have included A Night at the Opera, the Milky Way, Wild Animals, Winter Garden (based on the Russian Imperial gardens), Sea Treasures, and elaborate, abstract compositions from Byzantine to Baroque. Handbags have ranged from small jewellike shapes, which are sculpted like actual gemstones and crowned with monumental crystal clasps, to tactile suede and

Daniel Swarovski collection
MIROIR EVENING BAG. Autumn/winter 1994
Comet crystal stones
Crystal takes center stage. In the autumn 1994 collection, the Miroir bag perfectly expresses the metamorphosis of crystal.

silk evening bags encrusted with crystal having the appearance of waterfalls and icicles; there are even bejewelled pocket watches, and for autumn 1994, the ultimate interpretation, a basketlike handbag called *Le Miroir,* wholly constructed out of strands of mirrored crystal stones.

Similar ingenuity is used in the adaptation of several different techniques for decorating the handbags with crystal stones. Pictorial designs are hand embroidered with gold or silver threads, using the skills of the Paris atelier Montex, which works closely with Rosemarie Le Gallais. Great depth and softness are achieved by hand painting patterns on silk and velvet before adding the crystal embroidery. A particularly dramatic mosaic technique has been developed for sculptured purse handles, which has paved the way for a varied repertoire of shapes, colors, and textures.

Rosemarie Le Gallais has taken full advantage of the vast range of colors, shapes, textures, and special coating effects put at her disposal by Swarovski, and she spends a great deal of time in Wattens sorting through thousands of crystal stones. According to her theme and the season, she changes from clear crystal to a matte opaque finish, from pure, colorless crystal to rich colors, from faceted stones to smooth cabochons, and from classic lucid brilliance to the shimmering iridescence of specially coated crystal stones. She enjoys the way crystal can take on any guise, any color, any finish. Chameleon-like, it can, for example, be opaque chalk white to resemble porcelain or opaque orange-red to take on the characteristics of coral, it can be a dull metallic color, like an eighteenth-century mirror, or bruised with silver or bronze shadows.

In January 1992, the first Daniel Swarovski boutique opened in the rue Royale in Paris, a jewel box of a shop, designed by Roland Deleu to echo the collection's unique blend of traditional luxury with contemporary design. Louis XVI woodwork in natural oak is mixed with glass étagères. The boutique is the first of a series planned for the major capitals of the world to celebrate the Daniel Swarovski collection and to open a new chapter in the story of crystal.

Bořek Šipek for
Swarovski Selection
TABLE CLOCK. 1992
Crystal, metal. Height 8.5 cm,
width 12 cm, depth 9.3 cm
(3⅜ x 4¾ x 3¹¹⁄₁₆ in.)
The sundial form makes a witty
play on the importance of light
and shade in this simple but
powerful object.

Adi Stocker for
Swarovski Selection
FEDERHALTER
(PEN HOLDER.). 1992
Crystal, platinum-plated metal
Height 9.3 cm, width 13.2 cm,
depth 7 cm (3¹¹⁄₁₆ x 5⁵⁄₁₆ x 2¾ in.)
Inspired by the theater, the design
takes the form of a stage, with a
round green crystal globe at its
center. This could also double as
a photograph holder.

Meanwhile, as the Daniel Swarovski collection was coming to fruition, Swarovski was planning a new set of designer objects, aimed at a much wider audience. After four years of research and development, Swarovski Selection was unveiled in Spring 1992, a group of twelve cut crystal objects, more accessible and useful than the Daniel Swarovski objects, yet still at the cutting edge of contemporary design. The original collection was designed by a group of six avant-garde, up-and-coming young designers from France, Italy, and Austria: Giampiero Maria Bodino, Joël Desgrippes, Ludwig Redl, Bořek Šipek, Adi Stocker, and Martin Szekely. The international nature of the collection and the widely varied styles and backgrounds have given the project its intriguing and pioneering character.

Swarovski's brief to the designers was simply to take an innovative approach to cut crystal. No artistic restrictions were imposed, either on the design or on the function of the object. Today, the collection includes bowls, vases, and boxes as well as some surprising objects: a caviar bowl by Ludwig Redl, for example, a jewelry box by Adi Stocker, a table clock by Bořek Šipek, which is reminiscent of a sundial.

Always with an ear close to the ground, Swarovski has introduced new crystal objects and new designers, two per season, to the Selection range. From Sylvain Dubuisson comes an interesting business card holder, which plays with crystal's contradictory qualities, its transparency and solidity. The French designer Joël Desgrippes, famous for his perfume bottles, has taken the opportunity to branch out by creating a letter opener, which mixes smooth and clear cut crystal with opaque beechwood. Andrée Putman, an international leader in design, came up with two intriguing designs: a stalactite candle holder and a stalagmite ring holder. In this way, Selection provides sophisticated and diverse decorative objects, whose classical shapes take on modern and, sometimes, controversial twists.

The ongoing and progressive nature of Swarovski Selection, always picking up and interpreting in crystal the newest design trends around the world, has made this very much a living project, ensuring the continuing metamorphosis of crystal in the late twentieth century.

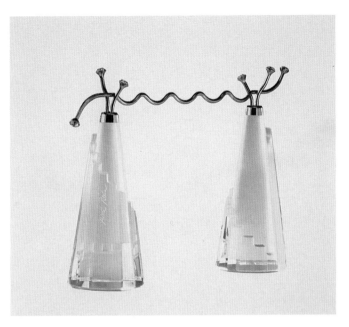

*Andrée Putman for
Swarovski Selection*
STALAGMITE RINGHOLDER
Fall 1994
Crystal, rhodium-plated metal
Height 11.4 cm, width 4.4 cm
(4½ x 1¾ in.)

Andrée Putman for Swarovski Selection. STALACTITE CANDLEHOLDER. Fall 1994
Crystal, chromium-plated metal. Height 16 cm, width 27 cm (6⁵⁄₁₆ x 10⅝ in.)

*Ludwig Redl for
Swarovski Selection*
KLEINE UHR MIT METALL-
SOCKEL (CRYSTAL CARRIAGE
CLOCK). 1992
Crystal, platinum-plated metal
Height 9.5 cm, width 6.7 cm,
depth 2.7 cm (3¾ x 2⅝ x 1¹⁄₁₆ in.)
*The clocks in the Selection line
are each fitted with a special
movement to set and change the
time at the touch of a finger.*

*Ludwig Redl for
Swarovski Selection*
KAVIARSCHALE
(CAVIAR BOWL). 1992
Crystal, platinum-plated metal
Height 9.3 cm, width 17 cm,
depth 11 cm (3¹¹⁄₁₆ x 6¹¹⁄₁₆ x 4⁵⁄₁₆
in.)
*A truly luxurious object whose
design cleverly balances crisp
geometric lines with soft curves.*

Joël Desgrippes for Swarovski Selection. Petit Vase (Small vase). 1992
Height 14.2 cm, width 12.8 cm, depth 8.3 cm (5⅝ x 5 x 3¼ in.)
*This extremely popular fan-shaped vase is designed as three separate parts, each showing a different aspect of the
versatility of crystal: flawless fluted transparency for the flower holder, a ring of opaque blue, and a full-cut base.*

*Giampiero Maria Bodino
for Swarovski Selection*
GRANDE CONTENITORE
(CRYSTAL BOWL). 1992
Height 7 cm, width 20.5 cm,
depth 11.8 cm (2¾ x 8⅛ x 4⅝ in.)
*A winged bowl, playing with
shape and light, that is both
humorous and elegant.*

INDEX
OF
PLATES

The following is a list of objects and the pages on which they appear in the book.

79
Adi Stocker
THE WOODPECKERS. 1988
Clear and matte crystal
Height 10.9 cm, width 6.65 cm
(4⁵⁄₁₆ x 2⅝ in.)

80
Michael Stamey
THE WHALES. 1992
Height 10 cm, length 9.3 cm,
depth 7 cm (4³⁄₁₆ x 3¹¹⁄₁₆ x 2¾ in.)

81
Michael Stamey
THE DOLPHINS. 1990
Height 7.8 cm, length 12.6 cm,
depth 7 cm (3¹⁄₁₆ x 5 x 2¾ in.)

81
Michael Stamey
THE SEALS. 1991
Height 5.3 cm, length 9.4 cm,
depth 7 cm (2⅛ x 3¹¹⁄₁₆ x 2¾ in.)

82
Gabriele Stamey
SANTA MARIA. 1991
Clear and matte crystal
Height 9.3 cm, length 11.4 cm
(3⅝ x 4½ in.)

83
Gabriele Stamey
OLDTIMER. 1989
Height 3 cm, length 8 cm
(1³⁄₁₆ x 3³⁄₁₆ in.)

84
Gabriele Stamey
FROG. 1994
Height 2.34 cm, length 3.23
cm, depth 2.26 cm (⅞ x 1¼ x
⅞ in.)

84
Adi Stocker
MINIATURE TOADSTOOL. 1989
Height 3.25 cm, width 3.2 cm
(1⁵⁄₁₆ x 1¼ in.)

85
Michael Stamey
THE ROSE. 1993
Length 8.1 cm (3³⁄₁₆ in.)

87
Claudia Schneiderbauer
HUMMINGBIRD. 1992
Clear and matte crystal
Height 7 cm, width 6.5 cm
(2¾ x 2⁹⁄₁₆ in.)

88
Claudia Schneiderbauer
BUMBLEBEE. 1992
Clear and matte crystal
Height 5.5 cm, length 4.6 cm
(2³⁄₁₆ x 1¹³⁄₁₆ in.)

89
Team design
LARGE BUTTERFLY. 1982
Crystal, gold-plated metal
Height 5.5 cm, width 5 cm
(2⅛ x 2 in.)

91
Max Schreck
EGG. 1979 (retired 1992)
Length 4.6 cm, diameter 6.3
cm (1¹³⁄₁₆ x 2½ in.)

91
Michael Stamey
SOUTH SEA SHELL. 1991
(retired 1994)
Height 4.7 cm, length 7.2 cm
(1⅞ x 2⅞ in.)

92
Anton Hirzinger
CENTENARY SWAN. 1995
Height 5.2 cm, depth 5.2 cm
(2¹⁄₁₆ x 2¹⁄₁₆ in.)

93
Adi Stocker
THE LION. 1995
Height 7.4 cm, length 12.9 cm,
depth 7 cm (2⅞ x 5¹⁄₁₆ x 2¾ in.)

95
Adi Stocker
EAGLE
See listing, page 7

96
Michael Stamey
CAT SITTING. 1991
Height 4.5 cm, length 4 cm
(1¾ x 1⁹⁄₁₆ in.)

96
Adi Stocker
MINIATURE DACHSHUND. 1987
Height 2.7 cm, length 5.3 cm
(1¹⁄₁₆ x 2⅛ in.)

BEAGLE PLAYING. 1993
Height 3 cm, width 3.4 cm
(1³⁄₁₆ x 1⅜ in.)

97
Max Schreck
GIANT OWL
See listing, page 2

98
Michael Stamey
APPLE. 1991
Height 6.2 cm, width 5 cm
(2⁷⁄₁₆ x 2 in.)

PEAR. 1991
Height 8.9 cm, width 5.9 cm
(3½ x 2⁵⁄₁₆ in.)

99
Max Schreck
Giant Pineapple. 1981
Crystal, gold-plated metal
Height 26 cm, diameter 17 cm
($10\frac{1}{4}$ x $6\frac{11}{16}$ in.)

100–101
Max Schreck
Chess Set. 1984
Clear and jet black crystal
Board 35 x 35 cm ($13\frac{3}{4}$ x $13\frac{3}{4}$ in.)

102
Adi Stocker
Aeroplane. 1990
Height 4 cm, length 7 cm
($1\frac{9}{16}$ x $2\frac{3}{4}$ in.)

103
Gabriele Stamey
Cathedral. 1990
(retired 1994)
Height 5.7 cm, length 3.4 cm
($2\frac{1}{4}$ x $1\frac{3}{8}$ in.)

103
Gabriele Stamey
Town Hall. 1993
(retired 1994)
Height 3.8 cm, length 6 cm
($1\frac{1}{2}$ x $2\frac{3}{8}$ in.)

Set of Houses (number 2)
1990 (retired 1994)
Height 3.4 cm, length 2.5 cm
($1\frac{3}{8}$ x 1 in.)

104–5
Gabriele Stamey
Train Set. 1988–93
Locomotive. 1988. Height 3.5 cm, length 6.5 cm, width 2.9 cm ($1\frac{3}{8}$ x $2\frac{9}{16}$ x $1\frac{1}{8}$ in.); Tender. 1988. Height 2.6 cm, length 2.6 cm ($1\frac{1}{16}$ x $1\frac{1}{16}$ in.); Tipping Wagon. 1993. Height 2.9 cm, length 3.9 cm ($1\frac{1}{8}$ x $1\frac{1}{2}$ in.); Petrol Wagon. 1990. Height 2.9 cm, length 3.9 cm ($1\frac{1}{8}$ x $1\frac{1}{2}$ in.)

106
Max Schreck
Large Pig. 1984
Crystal, rhodium-plated metal
Height 5 cm (2 in.)

106
Michael Stamey
Snail. 1986
Height 3.6 cm, length 4.3 cm
($1\frac{7}{16}$ x $1\frac{11}{16}$ in.)

107
Michael Stamey
Giant Duck. 1989
Height 11 cm, length 24.3 cm
($4\frac{3}{8}$ x $9\frac{9}{16}$ in.)

108
Max Schreck
Large Pyramid. 1976
(retired 1993)
Height: 6.4 cm, depth 4.8 cm
($2\frac{1}{2}$ x $1\frac{7}{8}$ in.)

109
Team design
Large Star Candleholder 1987
Height 11.25 cm, length 14 cm
($4\frac{7}{16}$ x $5\frac{1}{2}$ in.)

110
Martin Zendron
Lute. 1992
Height 7.9 cm, width 3.2 cm
($3\frac{1}{8}$ x $1\frac{1}{4}$ in.)

111
Martin Zendron
Harp. 1992
Height 9.98 cm, width 4.96 cm
($3\frac{7}{8}$ x $1\frac{15}{16}$ in.)

112
Michael Stamey
Sea horse. 1993
Clear and matte crystal
Height 8 cm, length 3.85 cm
($3\frac{3}{16}$ x $15\frac{1}{8}$ in.)

113
Michael Stamey
Shell with Pearl. 1988
Crystal, simulated pearl
Height 4.8 cm, length 5.88 cm
($1\frac{7}{8}$ x $2\frac{5}{16}$ in.)

113
Michael Stamey
Three South Sea Fish. 1993
Clear and matte crystal
Height 5 cm, length 8 cm
(2 x $3\frac{3}{16}$ in.)

114
Adi Stocker
Mother and Baby Panda 1994
Clear and jet black crystal
Mother: height 4 cm, length 4.5 cm, width 4.5 cm ($1\frac{9}{16}$ x $1\frac{3}{4}$ x $1\frac{3}{4}$ in.); Baby: height 1.8 cm, length 2 cm, width 1.4 cm ($\frac{11}{16}$ x $\frac{13}{16}$ x $\frac{9}{16}$ in.).

114
Adi Stocker
Polar Bear. 1986
Height 4.5 cm, length 8.8 cm
($1\frac{3}{4}$ x $3\frac{1}{2}$ in.)

115
Max Schreck
LARGE PENGUIN. 1984
Height 8.5 cm (3 3/8 in.)

116
Michael Stamey
THE KUDU. 1994
Clear and matte crystal
Height 10 cm, length 10 cm, depth 7.5 cm (3 15/16 x 3 15/16 x 2 15/16 in.)

117
Martin Zendron
THE ELEPHANT. 1993
Clear and matte crystal
Height 8.5 cm, length 11.8 cm, width 8.8 cm (3 3/8 x 4 5/8 x 3 1/2 in.)

119
Daniel Swarovski collection
MAYERLING GAUNTLET GLOVES. Autumn/winter 1989
Black suede kidskin, silk lining, hand embroidered

120
Ettore Sottsass for Daniel Swarovski
POSACENERE (ASHTRAY). 1989
Height 3 cm, diameter 18.5 cm (1 3/16 x 7 5/16 in.)

121
Ettore Sottsass for Daniel Swarovski
CENTROTAVOLA (CENTERPIECE). 1989
Clear and matte crystal
Height 22.1 cm, diameter 25.5 cm (8 11/16 x 10 1/16 in.)

122
Ettore Sottsass for Daniel Swarovski
VASO PICCOLO (SMALL VASE) 1989
Clear and matte crystal
Height 31 cm, diameter 7.6 cm (12 3/16 x 3 in.)

123
Ettore Sottsass for Daniel Swarovski
VASO GRANDE (LARGE VASE). 1989
Clear and matte crystal
Height 32.6 cm, diameter 18 cm (12 7/8 x 7 1/8 in.)

124
Ettore Sottsass for Daniel Swarovski
OGGETTO D'ARTE (OBJET D'ART)
Clear and red crystal. Height 27.8, width 18.5 cm (10 15/16 x 7 5/16 in.)

125
Ettore Sottsass for Daniel Swarovski
CANDELIERE (CANDLEHOLDER). 1989
Height 22 cm, width 8 cm (8 5/8 x 3 3/16 in.)

129
Alessandro Mendini for Daniel Swarovski
OGGETTO D'ARTE (OBJET D'ART). 1989
Clear and jet black crystal
Height 46.1 cm, diameter 18 cm (18 1/8 x 7 1/8 in.)

130
Alessandro Mendini for Daniel Swarovski
POSACENERE (ASHTRAY). 1989
Clear and jet black crystal
Height 8.2 cm, diameter 13.2 cm (3 1/4 x 5 3/16 in.)

130
Alessandro Mendini for Daniel Swarovski
CANDELIERE (CANDLEHOLDER). 1989
Clear and jet black crystal
Height 19.9 cm, width 8 cm (7 7/8 x 3 3/16 in.)

131
Alessandro Mendini for Daniel Swarovski
CENTROTAVOLA TROFEO (TROPHY CENTERPIECE). 1989
Clear and jet black crystal
Height 20 cm, diameter 13 cm (7 7/8 x 5 1/8 in.)

132
Stefano Ricci for Daniel Swarovski
CANDELIERE (CANDLEHOLDER). 1989
Height 17.1 cm, width 12 cm (6 3/4 x 4 3/4 in.)

132–33
Stefano Ricci for Daniel Swarovski
OGGETTO D'ARTE (OBJET D'ART). 1989
Height 25 cm, length 135 cm (9 7/8 x 53 1/8 in.)

136
Daniel Swarovski collection
BAVIÈRE BELT.
Autumn/winter 1989
Black suede kidskin, sterling
silver buckle, hand-embroi-
dered crystal mosaic

137
Daniel Swarovski collection
IMPERIAL EVENING BAG,
JARDINIÈRE LARGE AND
SMALL BROOCHES.
Autumn/winter 1993
BAG: black velvet, hand
embroidered with gold thread,
multicolored crystal stones;
BROOCHES: gold-plated ham-
mered metal, pâte de verre
leaves, multicolored crystal
stones

137
Daniel Swarovski collection
"IN TIME" EVENING BAG IN
THE FORM OF A POCKET
WATCH. Autumn/winter 1993
Black satin, hand embroidered

138
Daniel Swarovski collection
DIAMANT EVENING BAG
Spring/summer 1991
Red lizard skin, crystal clasp

139
Daniel Swarovski collection
"HANDLE" EVENING HAND-
BAG. Spring/summer 1993
Yellow satin, crystal mosaic

141
Daniel Swarovski collection
MIROIR EVENING BAG
Autumn/winter 1994
Comet crystal stones

143
Bořek Šipek for
Swarovski Selection
TABLE CLOCK. 1992
Crystal, metal. Height 8.5 cm,
width 12 cm, depth 9.3 cm
(3⅜ x 4¾ x 3¹¹⁄₁₆ in.)

143
Adi Stocker for
Swarovski Selection
FEDERHALTER
(PEN HOLDER.). 1992
Crystal, platinum-plated metal
Height 9.3 cm, width 13.2,
depth 7 cm (3¹¹⁄₁₆ x 5³⁄₁₆ x 2¾ in.)

145
Andrée Putman for
Swarovski Selection
STALAGMITE RINGHOLDER
Fall 1994
Crystal, rhodium-plated metal
Height 11.4 cm, width 4.4 cm
(4½ x 1¾ in.)

145
Andrée Putman for
Swarovski Selection
STALACTITE CANDLEHOLDER
Fall 1994
Crystal, chromium-plated
metal. Height 16 cm, width 27
cm (6⁵⁄₁₆ x 10⅝ in.)

146
Ludwig Redl for
Swarovski Selection
KAVIARSCHALE
(CAVIAR BOWL). 1992
Crystal, platinum-plated metal
Height 9.3 cm, width 17 cm,
depth 11 cm (3⁹⁄₁₆ x 6¹¹⁄₁₆ x 4⁹⁄₁₆ in.)

146
Ludwig Redl for
Swarovski Selection
KLEINE UHR MIT METALL-
SOCKEL (CRYSTAL CARRIAGE
CLOCK). 1992
Crystal, platinum-plated metal
Height 9.5 cm, width 6.7, depth
2.7 cm (3¾ x 2⅝ x 1¹⁄₁₆ in.)

147
Joël Desgrippes for
Swarovski Selection
PETIT VASE (SMALL VASE)
1992
Height 14.2 cm, width 12.8 cm,
depth 8.3 cm (5⅝ x 5 x 3¼ in.)

148–49
Giampiero Maria Bodino
for Swarovski Selection
GRANDE CONTENITORE
(CRYSTAL BOWL). 1992
Height 7 cm, width 20.5 cm,
depth 11.8 cm (2¾ x 8⅛ x 4⅝ in.)

INDEX
OF
ARTISTS

The following is a list of artists and the page numbers on which their works appear.

PHOTOGRAPH CREDITS

The publisher and author wish to thank those institutions and individuals for permitting the reproduction of works in their collections. Unless otherwise noted below, all photographs are by John Bigelow Taylor. References are to page numbers.

Luc de Champris: 29

Scott Hyde: 47

Michael Frank, Frankfurt/Main, Germany: 35 both, 120, 121, 122, 123, 124, 125, 129, 130 both, 131, 132, 132–33, 143 both, 145 both, 146 both, 147, 148–49

Foto Müller, Innsbruck, Austria: 14, 16, 19 both

Courtesy Swarovski: 10, 12, 22, 24, 37, 119, 136, 137 both, 138, 139